Comparative Agriculture

Hubert Cochet

Comparative Agriculture

Translation from French: Laurent Chauvet

Springer

Hubert Cochet
AgroParisTech
Paris, France

ISBN 978-94-017-9827-3 ISBN 978-94-017-9828-0 (eBook)
DOI 10.1007/978-94-017-9828-0
Springer Dordrecht Heidelberg New York London

Library of Congress Control Number: 2015934462

Éditions Quæ, R10, 78026 Versailles cedex, France www.quae.com

Printed on acid-free paper

Springer is part of Springer Science+Business Media (www.springer.com)

Contents

Part I Theoretical Approach to Comparative Agriculture

1 Comparative Agriculture, Object and Issues 3
1.1 Agricultural Development: Object of Comparative Agriculture ... 3
1.2 Issues Concerning Comparative Agriculture 6

2 The Origins of Comparative Agriculture and René Dumont's Legacy 9
2.1 Origins of Comparative Agriculture 9
2.2 Comparing in Order to Improve 10
2.3 Decisive Renewal for Agronomy and Agricultural Economics 12
2.4 Returning to the Field as an Antidote Against Theorising Drifts 14

3 "Agrarian System", Integrating Concept of Comparative Agriculture 19
3.1 Origin and Development of the Agrarian System Concept 19
Agrarian Structures or Agrarian Systems? The French Geographers' Input 19
The "Agrarian System" of French Agricultural Economists 22
3.2 Scales and Boundaries 28
From Agricultural District to Country, What Scale Should Be Privileged? 28
Spatial Bursting, Temporal Telescoping 31
The Agrarian System Goes Beyond and Encompasses the Sphere of Primary Production 35
3.3 Production System, Cropping System and Livestock Farming System 38
The Concept of Production System 38
Technical Choices Embedded in Social Relations 40

Production System Versus Activity System: A Badly Formulated Question ... 43
The Cropping System and Livestock System Concepts ... 45
3.4 Combination of Scales of Observation, Analysis and Comprehension ... 47
3.5 Between Life Sciences and Social Sciences, a Delicate Positioning of the "Agrarian System" ... 49

4 Diachronic Approach to Agrarian Systems ... 53
4.1 What is an "Agricultural Revolution"? ... 54
The Concept of "Agricultural Revolution" ... 54
The Four European Agricultural Revolutions ... 55
Agricultural Revolutions in Sub-Saharan Africa ... 56
Green Revolutions ... 61
Every Agricultural Disruption Is Not Automatically an "Agricultural Revolution" ... 62
4.2 Agricultural Crisis, Agrarian System Crisis ... 63
Crises Reflected in the Irreversible Degradation of the Environment ... 63
Endogenous Crises ... 64
External Crises ... 64
Political Crises? ... 65
Many-Sided Crises? ... 66
4.3 Comparative Agriculture, Agricultural "Crises" and "Revolutions" ... 67
4.4 Understanding Change in the Long Term to Better Identify the Process to Be Promoted ... 68

5 Comparing Productive Processes ... 71
5.1 Differences in Productivity on a Worldwide Scale and Consequences on Development ... 71
5.2 Which Forms of Agriculture Should Be Promoted? ... 74

Part II Methods and Expertise of Comparative Agriculture

6 Micro-Regional Approach to Agrarian Issues ... 89
6.1 Agricultural District: Privileged Object of Analysis in Terms of Agrarian System ... 89
6.2 Reading the Landscape ... 90

7 Terrain and Surveys ... 93
7.1 Conducting Surveys Oneself and Gathering Qualitative as well as Rigorously Quantified Data ... 93
7.2 Knowledge from the Top and Knowledge from the Bottom: Breaking Implicit Hierarchies ... 95
7.3 Tools, Machines and Skills: Agricultural Technology and Innovation Problematics ... 97

8 Making History in Comparative Agriculture 101
8.1 A Problem of Sources 101
8.2 Historical Surveys in Comparative Agriculture 103

9 How to Build Farm Typologies? 109
9.1 Brief Survey of Typological Methods 109
9.2 Prior Identification of Production Systems 110
9.3 Taking into Account the Conditions of Access to Land and Resources 112
9.4 Explaining Diversity 113

10 Economy of Agricultural Production Process 115
10.1 Linking Economic Approach and Technical Processes 115
10.2 Economy of Agricultural Production Applied at the Production System Level 116
Value Added and Productivity 116
Distribution of VA and Farm Income 120
Compared Economic Results: The Ukrainian Example 123
10.3 The Farmer as Homo-Oeconomicus 126
Homo-Oeconomicus 126
Cultural Factor and Tradition 128

11 Comparative Agriculture and Evaluation 131
11.1 Systemic Impact Assessment 132
11.2 Economic Assessment of Development Projects, from the General Interest Point of View 134

Conclusion 141

References 145

Introduction

Some of the major issues with which humanity is and will be confronted in the coming decades are the global agricultural and food issue. How to feed the planet? How should people *feed themselves* on the planet? And what productive processes should be promoted in moving progressively toward farming types that can ensure abundant and quality food production, preserving exploited ecosystems and their inhabitants, creating jobs and incomes, and contributing to reducing the gap in living standards that has been widening so much in the last decades? In trying to answer these questions and to light the way for decision makers as far as possible, it has become necessary to compare the many forms of farming found in each region or country of the world. But this comparison must not be seen through only by means of structural criteria made accessible to researchers through the statistical machinery. It must concern current processes, past and current trajectories as well as their methods of differentiation. It must be able to explain these evolutions and give them meaning. Finally, it must also concern their results as regards quantitative and qualitative production, the creation of wealth and income as well as the preservation or creation of jobs, and from the point of view of forms of ecosystem artificialization and their consequences.

This is the job of comparative agriculture which, as a discipline, was introduced at the Institut National Agronomique of Paris by René Dumont in particular, just after the Second World War, as a global and multidisciplinary approach to agriculture. Dumont was already pointing out the importance that should be given to the economic, social, and political conditions of agricultural development. From its very beginnings, and on the basis of this renewed approach to agriculture, comparative agriculture was gradually equipped with original concepts adapted to its purpose. It formed itself progressively into a fully fledged scientific approach.

Within the Social Science, Economics, and Management Department of AgroParisTech, today the *Comparative Agriculture and Agricultural Development* Unit trains agricultural economists to apprehend the historical and contemporary transformations of world agriculture so that they can formulate, manage, and assess agricultural development projects, programmes, and policies adapted to each situation. This research unit contributes also to training practitioners and researchers who are able to apprehend the technical, economic as well as social aspects of agricultural development, i.e., critical minds that are able to put technical issues back into the social context.

The aim of this book is to introduce this approach to agriculture, its concepts, and specific methods, on the basis of 25 years of experience in research and teaching, gained in various research fields in Latin America (Mexico, Central America, and the Andes), Africa (Burundi, Ethiopia, South Africa, Côte d'Ivoire, Guinea, and Sierra Leone), Asia to a lesser extent (Laos and Vietnam), and Europe (France and Ukraine); but also and especially on the basis of 20 years of team work within the *Comparative agriculture and Agricultural Development* Unit of the INA P-G, and later AgroParisTech. This book owes much to the discussions and field works conducted with my colleagues, and I would like to acknowledge them warmly here. The thoughts in this publication are personal as well as committed, sometimes provocative, always imperfect and incomplete, and certainly contradictory. I hope they can be extended and expanded into the future, thanks in particular to the reactions they will provoke.

This book is divided into two sections:

The first section is dedicated to theoretical thoughts on comparative agriculture. It presents at first the notion of "agricultural development," the very subject of comparative agriculture, but with a restored endogenous dimension, as will be seen in Chap. 1. We will then show, from the seminal work of Dumont as early as the middle of the twentieth century, how this approach to agriculture was slowly built up and consolidated, staying as close as possible to the field and keeping a certain distance from the theoretical approaches to development in vogue at the time, particularly just after the African Independences (Chap. 2). Long developments are then dedicated to the integrating concept of *agrarian system* around which comparative agriculture was built up (Chap. 3). The origin of this concept and its progress will be explored, as will the questions on the spatial scale of its application and the difficulties relating to the way it is used in certain situations. Its constituent subsystems, the production system in particular, will be analysed in detail, it being understood that it is indeed the combination of different scales of observation and analysis that give the agrarian system concept its efficiency. Chapter 4 deals with the diachronic approach to agrarian systems, and particularly with the importance of introducing the notions of "agricultural revolution" and "crisis," with a view to apprehending and comparing agrarian transformations in the long term. The synchronic approach to agrarian systems is tackled in Chap. 5 and, with it, the importance of the comparatist approach to production processes, their trajectories, and differentiation on a worldwide scale.

The second section is dedicated to the methods and expertise of comparative agriculture. Although this discipline is mostly the concern of agronomists and agricultural economists, and therefore specifically requires wide agronomic knowledge, it also calls on knowledge and skills from other social science disciplines. In this section, we will deal with the choice of tackling agrarian issues on a micro-regional scale, and the importance of being attentive when analysing the landscape underlying farming and breeding activities (Chap. 6). The survey methods implemented in comparative agriculture will then be presented (Chap. 7), as will the need to establish a real unbiased dialogue for the exchange of knowledge between scientists and farmers. Chapter 8 tackles the historical method peculiar to comparative agriculture and which, there again, gives more than is due to interviews conducted with actual producers. In Chap. 9, we tackle the issue of farm typologies which are essential to understanding reality, but which need to be handled tactfully. In this regard, we will insist on the importance of making the most of a prior reading of the landscape and a historical approach, so as to identify in a pertinent way the production systems to be studied. Chapter 10 is dedicated to the economic approach developed at the level of the production system. The economic results of the production system are understood as the results of its technical operation on the one hand (and as such are never analysed separately from it), and as the results of the methods used in accessing resources peculiar to each category of producers on the other, as well as the relative price conditions they fit into. The issue of assessment will then be tackled in Chap. 11. The expertise developed in comparative agriculture, may indeed prove to be efficient in identifying and measuring – as close as possible to reality – the differentiated impact of development projects and policies on producers and agrarian systems.

Part I
Theoretical Approach to Comparative Agriculture

Chapter 1
Comparative Agriculture, Object and Issues

1.1 Agricultural Development: Object of Comparative Agriculture

The expression "agricultural development" can be confusing, or at least offers several interpretations. According to the most widespread, development is a farming modernisation process, largely achieved thanks to the introduction and diffusion, by *extension agents*, of biological material and means of production stemming from research and the industry. In France, this conception was particularly fashionable during the Glorious Thirty, at a time when public or semi-public bodies, co-operatives or trade unions, represented locally by the SUAD[1] and CETA,[2] "were taking care" of development within the framework of the co-management of French agriculture,[3] established soon after the blueprint laws of 1960–1962. We could say that this general movement culminated during the Convention on Agricultural Development organised in the spring of 1982, soon after the victory of the Left, before declining thereafter. In the South, the multiplication of development operations and projects in countries actually described as "developing", within the framework of the green revolution in particular, related to the same conception of the word and subject, minus the co-management part . . .

Whether in the North or the South, it is the farmers' *participation* in development that quickly proved to be the *sine qua non* condition of development and, when there was no participation, the main limiting factor. If, as such, the participation of the populations was becoming the stumbling block of development, was it not because

---

[1]SUAD: Service d'Utilité Agricole et de Développement.

[2]CETA: Centre d'Experimentation des techniques Agricoles.

[3]The joint management of the agricultural sector by the authorities on the one hand, and by the agricultural professional organisations (majority farmers' association) on the other.

H. Cochet, *Comparative Agriculture*, DOI 10.1007/978-94-017-9828-0_1

it had first been conceived as something that came from elsewhere, from above, from outside? So much so that in its most restricted sense, ‘development’ came to designate only operations of agricultural extension, while *development sector – or extension service* – came to designate the socio-professional categories *in charge* of development: agronomists, technicians, extension agents etc. The objectives and the process were forgotten – all the more easily since it often took a long time for the effects of this development to be felt – in favour of the organisation in charge of development and its structure. Strangely enough, the meaning of the word came closer to the more remote meaning of ‘comparative agriculture’, and was sometimes employed in the field of marketing, as in ‘development: the packaging and presentation of a product for sale’ . . .

Considerable efforts were made, on the contrary, to try to bring research and development together into a common approach, particularly in the *research-development* and *research-action* approaches tried out from the 1980s onwards in various countries.[4] A testimony of this is the considerable scientific production on “development” in the countries of the North as well as the South, which will be impossible to analyse here.

The word “development” had a much wider meaning at a time when *development economics* was becoming specialised, as a scientific discipline, in the study of under-development and the means to overcome it. Economic theories of development – financing transition, developmentalist theses and dependency theories – caused a lot of ink to flow straight after the African Independences, before the specific fields of development economics (i.e. underdevelopment) became massively affected by neoclassic economics and its recent developments, aiming at giving an account of “market imperfections” and “informational dissymmetry”. The disenchantment that succeeded the enthusiasm and hope of the first post-independence years, the loss of credibility of the great paradigms of development and, in the end, the patent failure of so many development projects and programmes in the Third World, formed a favourable ground for radically contesting the very idea of development, understood as the new imperialism of Western thoughts, destroying local identities and cultures. After all, why should anyone desperately try to finance all sorts of projects if their results were ridiculous, not to say calamitous, when peasant farmers did not need any agronomists, economists, geographers and experts of all kinds in the first place? Has the “post-development” era begun, as is sometimes understood? Or should we turn the page of development[5] permanently?

While “agricultural development” is the privileged object of comparative agriculture, we still need to specify which of its concepts we will be adopting here,

---

[4]Guichaoua and Goussault (1993: 53–59). Many of these initiatives were typical of “development” agronomists directly involved in a compared agricultural approach or very close to it, for example on the Salagnac Plateau, in the Republic of Haiti, or on the hills of central Nepal (Bergeret and Deffontaines 1986). Concerning research-action, see also the publication of Lamballe and Castellanet (2003).

[5]As proposed unsubtly by Serge Latouche (2001).

which means that we need to discard all other remaining concepts (which some call "language remnants"[6]). There is the most common concept of 'agricultural development', which brings its process down to voluntary actions set by authorities, administrations or NGOs, and down to their effects. But we prefer the following, more comprehensive concept: *A general long term agricultural transformation process of which the elements, causes and mechanisms can be of endogenous origin and, at the same time, the fruit of different exogenous inputs, enrichments or innovations*. This concept is obviously far richer and more complex than all the desired or real effects resulting from projects, programmes or policies established to try to modify the meaning of agricultural development, which is a fortiori much vaster than the mere flow charts of the so-called "development" (or extension) services.

As an example, all agricultural transformations characterising the agricultural revolution of the eighteenth and nineteenth centuries in Western Europe (e.g. cultivation of fallow lands with fodder roots and tubers and with forage crops, development and intensification of animal farming, increase in cereal yields) were an authentic development process, as understood in comparative agriculture. Similarly, the changes taking place in the agriculture of Burundi and Rwanda during the eighteenth century, i.e. well before colonisation, with the generalisation of American plants, the modification of cropping systems and disruption of farmers' work plan, the spreading of harvestings and substantial feeding improvement, and the doubling of labour productivity among others, also formed a true *agricultural development* process, in the full sense of the word (*infra*).

Without for all that grasping the importance of the abovementioned transformations which, owing to their scale, constitute what we call an agricultural revolution (*infra*), "agricultural development can be defined as a progressive change in the agricultural production process, towards an improvement of the cultivated environment, tools, biological material (cultivated plants and domestic animals), working conditions on the farm and the satisfaction of social needs" (Mazoyer 1987). This is indeed the *object* of comparative agriculture. Whether this process can be considered as progressive, in the first sense of the word, or on the contrary reflects the degradation of its constituents is another matter. And it is precisely on the basis of a detailed study of these *development* processes and aspects which are progressive as well as regressive and contradictory, that we can take a critical look at what has been done in the past and is currently being done as regards "development" (in the restricted sense of the word this time), and that new proposals can be outlined. Studying and measuring the impact of development projects and agricultural policies on the agricultural sector, is perfectly suited in comparative agriculture.

[6]This expression is from Philippe Couty and actually concerned the word "development" (1981).

1.2 Issues Concerning Comparative Agriculture

Now that we have a clearer idea of its object, it will be easier to define comparative agriculture and its mains issues. According to Marc Dufumier, the idea is to "understand agricultural realities so as to modify agricultural development" (1996a), a definition that encompasses the two dimensions of comparative agriculture, i.e. the cognitive dimension with the imperative of producing knowledge in order to better understand ongoing processes, and the applied dimension with the main objective of contributing to the elaboration of development projects, programmes and policies likely to modify the course of agricultural development for the common good. Elaborating on these two foundations of the discipline, Marc Dufumier wrote, on the one hand: comparative agriculture "aims to make historical processes intelligible, processes through which the various worldwide agrarian systems have been brought to evolve, under the double dependency of ecological conditions and socioeconomic transformations. It presents and develops the theoretical reference framework that makes it possible to put each one of the specific realities or agrarian situations back in their historical perspective, in relation to and in comparison with the more general movement of differentiation of agrarian systems in the world" (1996b: 303). He specified on the other hand, that the main imperative of comparative agriculture is to "conceive the new agro-ecological and socioeconomic conditions to be created, so that the different types of farmers have the means to implement production systems complying the most with the general interest, and that they also find it worth their while. This supposes to be relatively well-informed about agro-ecological and socioeconomic elements on which it is appropriate to intervene first and foremost to modify the behaviour of farmers and the future of their production systems" (1996a: 927). In the 1989 edition of the *Grand Larousse Universel*, Marcel Mazoyer also wrote, under "comparative agriculture": "comparative agriculture makes every effort to discover the conditions for a development which is adapted to each situation and viable, i.e. which can be reproduced", in other words a "sustainable" agricultural development, as will be seen later on, although this expression was not yet used at the time.

Comparative agriculture endeavours to build "a body of global knowledge that explains the origins, the transformations and the role of agriculture in the evolution of man and life, in different eras and parts of the world, a body of knowledge that can become integrated into general culture and, at the same time, constitute a conceptual, theoretical and methodical basis, for all those who ambition to intervene in agricultural, economic and social development" (Mazoyer and Roudart 1997).

As such, comparative agriculture is the science of transformations and adaptations of agricultural development processes. Yet, faced with the extraordinary diversity of agricultures around the world, how can we appreciate the mechanisms, processes, regulating modes and contradictions of each one of them? And how can we establish a meaningful comparatist approach, i.e. leading to a better understanding and perspective of each specific type of agriculture and, at the same time, to an informed perception of the modes and consequences of their interlinking, under the growing influence of the markets in particular? The scientific study of the most unusual agrarian system can only find meaning when confronted with other agrarian systems. "Whether this study seeks to show the absolute specificity of a population or place [. . .], or its objective is to show that this population or place is part of a vaster whole [. . .], it will always refer to the rest of the world to assert differences or similarities".[7] "However, in all cases, showing the uniqueness of a scientific object depends on its distinction from other cases. The identity of a scientific object is linked to its singularity and, at the same time, to its difference."[8]

What should be compared and why? Should we compare agrarian societies that are "similar" or, on the contrary, should we privilege the comparison of agricultures as different as possible? This process would be endless and without purpose. That is why the comparatist approach cannot be limited to make inventories of similarities and differences. What needs to be compared are the processes more than the actual objects. It is in this light that comparative agriculture seeks:

(1) to identify what is universal or, on the contrary, unique, what seems to be fundamental or rather secondary in the organisation of agricultures and their dynamics;
(2) to determine, interpret and explain these differences by "resituating each specific situation within the more general framework of the differential evolutions of agriculture on a worldwide scale" (Dufumier 2002b: 68);
(3) to emphasise continuities and/or breaks, relationships, progressive series, one or several overall dynamics;
(4) to retain from this agrarian legacy of humanity "ways of doing" and know-how, tools, mechanics and machines, ideas, vegetal and animal material, in short, everything that can contribute to enlighten, direct or favour, in a specific situation, agricultural development in a way which conforms more with the common good.

Yet, the comparatist analysis, as enriching as it is, is not sufficient on its own, unless it is equipped with tools and concepts to organise the extraordinary diversity of situations and give it meaning. "Stricto sensu, the comparative analysis can only reproduce the diversity of observable phenomena without reducing it and without deriving even slightly general laws that can explain these phenomena" (Mazoyer 1974). Moreover, "how to make the diversity of concrete forms taken on today by agriculture across the world intelligible, and learn general lessons from it, without for all that ending up with abusive generalisations or modellings that are too simplifying?" (Dufumier 2002b: 62).

It is in this perspective that, for the past 40 years or so, comparative agriculture has built up its own theoretical concepts and developments, on the historical evolution and geographic differentiation of agrarian systems. Before analysing the concepts developed by this discipline as well as the investigation methods established to define its object, we need to look at the origin and history of comparative agriculture.

---

[7] Philippe Gervais Lambony, concerning comparison in the social sciences (2003: 3).

[8] Idem (p. 33).

Chapter 2
The Origins of Comparative Agriculture and René Dumont's Legacy

2.1 Origins of Comparative Agriculture

François Sigaut, Director of Research at the Ecole des Hautes Etudes en Sciences Sociales, recalled recently that Dumont did not invent comparative agriculture, but reinvented it: "When he began his career, we had practically forgotten the name and the subject, except for the corresponding chair of agronomy . . ." (Sigaut 2004: 17).

Indeed, it would seem that we owe the appearance of experimental agronomy and that of comparative agriculture to Duhamel de Monceau (1700–1782). The establishment of experimentations (concerning the use of mechanical seeders and horse hoes), under different and partly uncontrollable circumstances, made careful observation and the *comparison* of practices necessary.

One century later, and while the perspectives of progress were widening, reasoning practices out was no longer on the agenda, but changing them as quickly as possible was: "Agronomists then stopped taking an interest in practices – these were left to folklorists – to become progress extentionists. The past, and even the present insofar as it is only considered as an extension of the past, were no longer important. Only the future mattered, a future that was no longer decided in the field but in laboratories and research stations" (idem).

François Sigaut continues: "René Dumont is probably the first person, in the 20th century, who fought against this mistake, the consequences of which have not yet been measured" (idem). Thus, Dumont's lesson and his reinvention of comparative agriculture were about returning to observations and comparatist analyses.

Yet, the "Chair of Comparative agriculture" was founded as early as 1878 by Eugène Risler, who proposed a course on "The comparison of agricultural systems currently practiced in different countries, under various economic, geological and climatic conditions" (Boulaine and Legros 1998: 219). The incomparable merit of his major work entitled *Géologie Agricole* (Risler 1897), was that it insisted on the spatial diversity of agriculture, which many experts had neglected before him. Despite the title which reveals the author's point of view and expresses a certain

H. Cochet, *Comparative Agriculture*, DOI 10.1007/978-94-017-9828-0_2

determinism as far as the physical environment is concerned, Risler examined also agricultural practices and the characteristics of livestock farming: "in order to better understand these countries, he becomes a sociologist and sometimes a historian; he is also an economist (...) and foresees economic globalisation; he discusses customs duties and protectionism..." (Boulaine and Legros, op. cit.: 222).

However, it was Dumont who gave this global and multidisciplinary approach its full dimension, by emphasising the importance of economic, social and political conditions to describe and understand the multiple forms and paths of agricultural development (Dufumier 2002a). As an "agronomist concerned with hunger around the world", Dumont was particularly concerned with increasing the food production of the countries the most affected by malnutrition. Nonetheless, he had to face the facts: "Technical progress in agriculture cannot be reasoned out from outside. It is part of a complex network of social relations (right of access to land, distribution of production means, available savings for the acquisition of new means of production deemed 'modern', the place of the different categories of farmers in the social process of exchange and redistribution of the fruits of labour, etc.). As such, the centre of gravity of the agronomist's preoccupations is shifting progressively, from technical expertise towards understanding the social relations of production, predicting the implementation of this expertise" (Kroll 1992: 10).

2.2 Comparing in Order to Improve

Dumont's only published text dealing explicitly with comparative agriculture, seems to be that published in the 1952 edition of the *Larousse Agricole*, under the article entitled "Comparative agriculture". As a Senior Lecturer at the Institut National Agronomique, he then outlined the "definition and basics" of comparative agriculture: "Comparative agriculture proposes to study the essential characteristics of agriculture in various geographic units (hamlets or suburbs, *communes*, *cantons*, areas, regions, nations or continents) in order to try to find means of improvement. [...]It seems that, as part of the agronomist's work, his main duty is to recommend not modern techniques [...] but the choice of the most recommended animal and vegetal operations, as well as the *framework* in which the application of these techniques would be less difficult" (Dumont 1952).

Comparing in order to improve, orientate and develop agriculture: this is what, in Dumont's mind, seemed to be the first objective of comparative agriculture and its *raison d'être*. Solicited from all quarters to give his opinion and involved in agricultural planning soon after the Second World War, Dumont outlined the major production areas of the future, selected for each region activities worth developing, i.e. those that seemingly would have been condemned in the future. Comparative agriculture was, on a worldwide scale, the approach that helped him move forward. At the time, it is true, French agriculture, despite important regional contrasts stemming in particular from the first regional specialisation movement enabled by the development of the railway at the end of the nineteenth century, was still largely dominated by mixed crop-livestock farming systems, combining a large number of productions, and as such was little specialised. Dumont explained: "After describing existing cropping and livestock farming systems, the role of the agronomist is therefore to advise the farmer (on sole proprietorship economics), otherwise sometimes to take a decision (on programme economics): in any case, to indicate the desirable evolution of these systems. This is the most delicate task that can be requested of him. Elements of knowledge on the environment [. . .] and the current production systems must be tied up with the national objectives of the programme that indicates those speculations to be developed and those to be slowed down" (1952: 907). At the time, the idea was to boost a movement of regional specialisation deemed in keeping with the common good (which is the meaning Dumont always gave to planning), a movement that was to be conducted according to the comparative advantages of each region . . . "In a context deeply marked by 70 years of Melinist protectionism,[1] by the celebration of farming autarchy by conservative Republicans as well as Agrarian Catholics, this call to trade, exchange and specialisation is an unequalled break" (Hervieu 2002).

By combining the approach to comparative agriculture with the need for regional and worldwide specialisation, Dumont made of this discipline a science that became fully geared towards action and, in the context of the time, towards the establishment of the broad outlines of agricultural specialisation. As a result, he proposed the following: giving up cereals and developing livestock farming systems in the mountains with an increase in forage supplies; ploughing permanent meadows and intensifying forage production in the West and on the margins of the Massif Central; dispersing vineyards and increasing quality; replacing lands that were still fallow on the limestone plateaux of the North–East with temporary meadows and intensifying livestock farming; decongesting Armorican bocages, uprooting hedges and intensifying dairy production in the West; irrigating and producing hybrid maize and associated livestock in the South–West; irrigating the Lower Durance and developing fruit tree cultivation; developing colza as rotation head in the Paris Basin to compete with African groundnut, but giving up sugar beet, because competition with sugar cane was too harsh, among other things.

Recalling the development of heavy long-distance transportation at the end of the nineteenth century, and its consequences on the first specialisation movement, he declared curiously: "And so we have entered the *dynamic phase of comparative agriculture* before the 20th century", almost likening the discipline to the result of its application (Dumont 1952: 904). Dumont even saw in the beginnings of future production quotas (of beetroot) and *appellation contrôlée* wines, monopolistic tendencies leading to a status quo, which would make any work on comparative agriculture useless (idem: 926)!

---

[1]Méline (Jules), French Minister of Agriculture under the Third Republic, then Prime Minister, the main instigator of the protectionist policy of the time.

Comparing in order to improve: comparative agriculture was not yet a scientific discipline per se but, rather, an action science. He added in this regard: "In reality, the subject is too complex and insufficiently studied to be able to bring out universal laws; it remains an art that will rather indicate tendencies, and sometimes even fairly general rules. An art that will require a lot of commonsense for the simultaneous implementation of extraordinarily multiple scientific data, which any regional agricultural study must rely on" (idem: 903).

2.3 Decisive Renewal for Agronomy and Agricultural Economics

Beyond the objectives of comparative agriculture mentioned by Dumont at the time, which were slightly restricted because of the specific post-war context, the fact remains that the comparison of farming systems established in different parts of the world, brought a new dimension to agronomy in the general sense of the word. Widening considerably the comparatist perspectives initiated by Arthur Young[2] two centuries before that, Dumont began a comparison of the different farming systems worldwide.

For example, the *comparison* of gross labour productivities, measured in kilogrammes of cereals produced per work day dedicated to cultivation, a comparison systematically put forward in Dumont's works[3] at the time already, proved to be increasingly unavoidable in trying to foresee and anticipate the respective evolutions of the agricultural sector, in different regions of the world. Dumont was the first to point out the incredible gaps in labour productivity, and to inquire about the huge consequences of these inequalities worldwide. With a hindsight of more than half a century on these works, today while the increasing unification of the market and prices, and its consequences, are mobilising international conferences and large-scale social organisations, we can appreciate the visionary nature of these works and, at the same time, the necessity of measuring and understanding these gaps in productivity, which have in fact increased tenfold since Dumont's first works. The idea that the world is *one* and that what happens in a given place cannot be truly understood without referring to what is happening on the other side of the planet, thereby already outdating any approach to agronomy or geography by geographic field, from then on constituted the main thread of comparative agriculture.

The fact that the observation and description of farmers' practices is making a big comeback, undoubtedly also constitutes a founding act of comparative agriculture. "In any case, comparative agriculture must study cropping and livestock farming

[2]Arthur Young (1741–1820) is an English agronomist and economist, the author of *Travels in the Kingdom of France*, published just after the French Revolution (1792).

[3]Including in this first definition of comparative agriculture written for the *Larousse Agricole* of 1952. See also *Economie agricole dans le monde* (1954) and Marc Dufumier (2002b).

systems in their current state, but also from a dynamic perspective, by following past and especially recent changes, which indicate the direction of the evolution" (Dumont 1952: 907). Specific knowledge of the environment, the current state of agriculture and the direction of its historical evolution, needs to be accompanied by an inventory of available human and material resources, then by current and potential prospects. In the "essential characteristics" attributed to comparative agriculture, Dumont outlined "what a study of comparative agriculture [ought to] describe": climatic and soil conditions, human conditions and land tenure, the dimension of production units and their source of energy, cropping and livestock farming systems as well as their dynamic, the degree of intensification and equipment (idem: 904).

The first 30 years in the professional life of Dumont (1933–1961) were marked particularly by long field studies, as well as several works relating in minute detail the main characteristics of agriculture in a few large regions of the world (Tonkin, United States, France and China[4]). On reading these works, one is struck by the way in which observations and conversations with farmers are reproduced. His works did not include any standardised survey or sampling guide, but loads of observations organised according to a path suggested by intuition, and carefully selected in the agricultural diversity of the region studied, sometimes also imposed by chance encounters: an abundance of concrete images, testimonies and meticulous survey reports, documenting with frightening accuracy crop sequences and crop management sequences, herd flock movements and forage calendars, the description of ploughing equipment and work organisation, productivity and farm incomes among others. In the Belgian colonies, it was Pierre de Schlippé who initiated this return to the field, direct observation and survey, using his own method of "agricultural anthropology" (de Schlippé 1956a, b).

Because he gave as much importance to the detailed and accurate description of Indochinese or African farmers' actions as those of French farmers in this demanding comparison, Dumont also ended up shaking colonial agronomy rather abruptly, although beneficially. At the time, only ethnologists for the most part could afford to spend time observing and describing in detail the actions of distant populations. When Dumont was sent to Tonkin, it was to "improve" Indochinese rice-growing; but when he studied traditional rice-growing techniques at length, in the end he concluded that they were "generally valid" (Dumont 1954: 15). Making of a technical sequence of slash-and-burn agriculture an object of study as important and interesting as a three-field system in Eastern France, attributing a certain "validity" to it and measuring its labour productivity with the same tools and methods, is what sparked new and long-lasting interest among agronomists in what was then called the *Third World*. At the time of the Independences and in the new abundance of ideas to which they gave rise, many social sciences were shaken

---

[4]*La culture du riz dans le delta du Tonkin* (Société d'éditions géographiques, maritimes and coloniales, Paris, 1935), *Les leçons de l'Agriculture américaine* (Flammarion, éditeurs, 1949), *Voyages en France d'un agronome* (Editions M.-Th Génin, Paris, 1951), *Révolution dans les campagnes chinoises* (Editions du Seuil, Paris, 1957).

and, as a result, compelled to renew or "reposition" themselves. As a technical science, agronomy escaped this renewal of ideas for many years still, without the comparative agriculture of a Dumont or other agronomists who, incidentally, were few in numbers.[5]

Breaking away from teaching the "special cultures" of our colonies in favour of a class on comparative agriculture,[6] asserting the validity of the farming practices of the "autochthons", suggesting that certain North American farmers or Algerian settlers had been deteriorating the national land because of their erosive practices, as much as or more than their African colleagues who were practicing slash-and-burn agriculture, all this in 1952 represented a real challenge, all the more daring since it was taking shape inside a prestigious Paris-based high education institute....

2.4 Returning to the Field as an Antidote Against Theorising Drifts

At the time when the development of comparative agriculture was spurred on by Dumont, then by Mazoyer from 1974, the world was experiencing major political evolutions, on the occasion of the African Independences in particular. The debates at the time on development and underdevelopment, gave rise to the tremendous expansion of the social sciences around the notion and interpretation of "underdevelopment" (Guichaoua and Goussault 1993). The major schools of thought of the time (theories of modernisation, dependentist approaches, Marxist approaches, etc.) could not ignore the agricultural sector of the societies they were trying to understand. Their influence on comparative agriculture could have been notable, particularly the eminently modern idea according to which the economies of the South and the North are interdependent on the world market, or that according to which underdevelopment in the South could not be understood or analysed, unless it was integrated as an endogenous component of the development of the countries of the North.

In the Anglo-Saxon academic world, that era was marked by the multiplication of research works claiming to be *peasant studies*, marked at the beginning in particular by Shanin (1970), later to be outlined by Bernstein and Byres (2001). In the context of studies on the development of poor countries in the days following the Independences, the idea was to see how knowing more about these farming communities could lead to apprehending better pre-capitalist agrarian formations in different parts of the world, the paths of developing countries transiting towards capitalism, as well as the consequences of colonial development on the dynamics

[5]In Belgium, it was undoubtedly Pierre de Schlippé who assumed the break away from colonial agronomy.

[6]This is what Dumont proposed in 1953 in his note "title and works"... written for his presentation at the competitive exams to become a teacher...

and processes of the subsequent development/underdevelopment (Bernstein and Byres 2001). At a time when Marxist references occupied an important place in the social sciences, sound debates concerned (1) pre-capitalist formations and "feudal production mode", (2) "transition to capitalism", (3) "transition to socialism", (4) colonialism and (5) the relationship between development and underdevelopment.

Fundamental inputs from the development economics of the time undoubtedly include, on the one hand, the idea of the *deterioration of the terms of trade*, therefore calling into question the importance of the countries of the South specialising in primary (particularly agricultural) products and, on the other hand, the notion of *dualism*, which postulated the existence of a structural surplus of labour in the economies of the third world (Assidon 2000). The first fuelled a long and animated debate on export crops versus food crops, which has become slightly outdated today; however, the basic idea behind it, i.e. the relative decline in the remuneration of producers, is more than ever topical. As to the idea of dualism, it inspired policies "to put the labour surplus to work" (set to be transformed into profit), in the domain of agriculture in particular, with an authoritarianism and dogmatism that were just as bad as the development policies of the colonial era.

One must say that, at the time, all the theoreticians of underdevelopment were almost in complete ignorance of peasant farming, and displayed little regard for the farming community. Bogged down in so-called "pre-capitalist" social relations, or open only to *kulakisation*, these farmers were set to "evolve" in accordance with the principles in force, i.e. "modernisation", transition towards capitalism or collectivisation. What was the use of studying their practices when these were doomed to a rapid and desirable disappearance? From West to East, there was a common refusal to take into account the heterogeneity of the national situations and concrete processes of social and economic change (Guichaoua and Goussault, op. cit.).

That era was also marked by another theoretical debate animated around the issue of agricultural development in particular: the debate that saw Malthusians and Neo-Malthusians opposing followers of the theory of Ester Boserup. Countless works on the development of African agriculture in particular, recalled in their introduction the theories developed in their days by Thomas Robert Malthus and Ester Boserup, as if the definition of a research problematics could not do without these references, or established one of the two theories as an interpretative model of the transformations observed.

In England, at the end of the eighteenth century, despite the considerable ongoing changes in the English agricultural revolution, Malthus[7] made of the food production level (his "subsistence level") an independent variable, or one that could only be extended in proportion and according to the extension of cultivated surface areas, to the detriment of the unlimited forests or lands of America. Under these conditions, the geometric increase in population inevitably resulted in the two curves

[7]Here, we are referring to An Essay on the Principle of Population published in 1798, which subsequently inspired his more famous text from 1803.

of subsistence and population crossing each other, and in the triggering of a cycle of "Malthusian regulation": a preventive brake for the wealthy classes haunted by the thought of moving down the social scale (pushing back marriage timelines, waiting longer before having a second child, etc.), starvation, destitution and infant mortality for commoners. But destitution had its virtues: by provoking salary cuts, it encouraged ploughmen to hire more hands to increase clearings in particular and, as such, to expand cultivated areas, which raised the subsistence level, encouraged in the process the recovery of the birth rate, and met once more all the requirements for a new "cycle of regulation". Pessimistic and, at the same time, fatalistic, steadfast in his belief in the unchanging order of things and social order, Malthus did not envisage any form of technical – and even less – social progress.

Almost two centuries later and in a very different historical and demographic context, developing countries experienced their first demographic transition, which Neo-Malthusian approaches explained with the deterioration of living and production conditions resulting from the population explosion. Unless emigration was conceivable and the colonisation of new lands possible, population densification was necessarily leading to the over-exploitation of lands, the diminution of their fertility and the accelerated degradation of the environment.[8]

More than 150 years after Malthus and in reaction to Neo-Malthusian currents, Ester Boserup published in 1965 *The Conditions of Agricultural Growth, The Economics of Agrarian Change under Population Pressure*. In it, demographic growth was becoming the independent variable. By provoking an underlying drop in hourly labour productivity, and by compelling people to change their production techniques, demographic growth was then becoming the true driving force of agricultural progress. Published right in the middle of a population explosion, and by relying on a gradual series that was supposedly characteristic of developing countries (the progressive reduction cycle of the duration of the "fallow", by changing successively from long-term forest fallows to intensive cropping systems with several cycles per year, via all intermediary stages), this text brought a breath of optimism, and somewhat rehabilitated farming societies by endowing them with an endogenous capacity to evolve and modernise. Increase in cropping rates, technical change and intensification... but under what conditions? Two were sufficient for the author: hourly productivity had to be dropping, which is what "motivates" farmers in changing techniques, and additional work had to be generated collectively and dedicated to agriculture, particularly in the form of "labour investment"[9] for land developments, as required when changing from one technique to another (marsh drainage, erection of terraces, hydraulic infrastructures, etc.).

Finally, whether inspired by Malthus or seduced by the hypotheses of Boserup, everybody agreed on what seemed essential, i.e. that it was indeed the popula-

[8]Concerning this debate opposing, during the 1950s, Malthusians and "optimists", see Sauvy (1958).

[9]"Investment" or "rural investment" in E. Boserup's work, although the author refers to an investment as labour more than capital.

tion/resource relationship that was the driving force behind agrarian dynamics. Combined in the greatest simplicity, these theoretical references in a way exempted anyone from having to look further, and to think about the true nature of crises and how to confront them.

Beyond the theoretical debates of the time, which very often flouted the specific characteristics of each situation, the heterogeneity of the national situations and concrete processes of social and economic change represented the actual object of Dumont's investigations. Although during his "productivist" and "developmentalist" period, he might have justified and appropriated certain aspects of these policies "to put the peasantry to work", his constant field work made him stay in the background in relation to these major schools of thought. And it was thanks to this distance, maintained de facto through comparative agriculture, with these major schools of thought and global interpretations of underdevelopment in the 1960s and 1970s, that Dumont was actually able, and the first, to anticipate the failures of development and to denounce them, not without causing a certain sensation, in *L'Afrique noire est mal partie* (1962).

Later on, and while the major development models (and schools of thought) began to fall apart together with the disillusions of development, the fall of the Eastern Block and the general increase in development inequalities, operational pragmatism had become essential in the same social science disciplines (Guichaoua and Goussault, op. cit.), very much like an approach that had already been established by comparative agriculture. Because its object of study lent itself to concrete and localised field studies more than global constructions, comparative agriculture went through this update without any crisis, unlike other disciplines or schools of thought developed soon after the Independences.

Since then, a new global steamroller has been threatening the social sciences: liberal globalisation with an economic theory supposed to be unique and capable, on its own, of explaining the present, as well as anticipating and prescribing the desirable transformations of societies, including the agricultural sector. Once more, it is the return to the field, to the *local*, to the meticulous search for *concrete* dynamics of development or marginalisation in different regions of the world and the *comparison* of current processes, which is more in a position to go beyond simplifying or distorting explanations elaborated away from the field.

Chapter 3
"Agrarian System", Integrating Concept of Comparative Agriculture

The following pages are dedicated to the concept of "agrarian system". After recalling the origin of the concept and its recent developments within the framework of comparative agriculture, we analyse the issue of scale of analysis to be privileged and its boundaries, and with it, that of the difficulties potentially arising when using this concept. In a third section, we endeavour to explain in detail the sub-systems making up the agrarian system, i.e. first of all the *production system* (at the level of the basic production unit or farm), and secondly the *cropping system* (at the level of the plot), and its counterpart the *livestock farming system* (at the equivalent level of the herd). The fact that these scales of observation, analysis and understanding need to be combined is addressed subsequently. Finally, we will question the atypical status and position of this concept between life sciences and social sciences.

3.1 Origin and Development of the Agrarian System Concept

Agrarian Structures or Agrarian Systems? The French Geographers' Input

Geographers were the first to speak of "agrarian system", and the first definition of this concept was probably due to André Cholley (1946). Concerning the research method adopted in rural geography, he wrote: "we could manage to study reality more closely by considering that agricultural activities reveal a real combination or set of elements borrowed from different yet very closely related domains; elements so interdependent that it is not conceivable that one of them is radically transformed without the others being considerably affected, and that the entire combination is not, as a result, modified as far as its structure, its dynamism and even its external aspects are concerned" (Cholley 1946: 82).

H. Cochet, *Comparative Agriculture*, DOI 10.1007/978-94-017-9828-0_3

Twenty years later, the thesis of Paul Pélissier on *Les Paysans du Sénégal* (The Farmers of Senegal), offered a remarkable example of global and systemic approach to agrarian societies, although the agrarian system concept was not brought up (Pélissier 1966). The article published in 1964 by Pélissier (and co-authored with Gilles Sautter), entitled *Pour un atlas des terroirs africains: structure-type d'une étude de terroir* (Towards an Atlas of African *Terroirs*: Template for a *Terroir* Case Study), was behind a remarkable series of studies on African *terroirs*, conducted by various researchers under the supervision of these two authors.

Thirty years after Cholley's article, it was not so much the *agrarian system* that Georges Bertrand revived in the *Histoire de la France Rurale* (The History of Rural France), but the *agrosystem*: "Each agrosystem corresponds to a certain relation between a type of rural society and a type of environment, at the material level as well as that of behaviours and mentalities" (Bertrand 1975).

Straight after the publication of *Histoire de la France Rurale*, Claude and Georges Bertrand even proposed to study landscapes *as* a system (1978): "The simplest and most ordinary landscape is social and at the same time natural, subjective and objective, spatial and temporal, it is a material and cultural production, real and symbolic, among others. Counting and analysing separately the constituents and the different spatial, psychological, economic and ecological characteristics among others, do not lead to managing the whole. The complexity of the landscape is morphological (form), constitutional (structure) and at the same time functional and one must not seek to reduce it by dividing it . . . . The landscape is a system."

More recently, another geographer, J. Renard, wrote: "An agrarian landscape is also, over a vaster space and on a smaller scale, a series of repetitive elements being laid out in the same way and making up an agrarian structure with, as a result, an agrarian combination which, concretely, is the agrarian morphology. (. . .)In the background, there appears clearly the notion of agrarian model or type, i.e. beyond the diversity and complexity of the forms, the recognition of regularity, order and organisation. This is evidence that the agrarian landscape which is a layout of places, is indeed and first of all a social event" (Renard 2002: 13).

However, in the mind of most French rural geographers, and despite the remarkable and premature opening of A. Cholley on the matter, the expression "agrarian system" was used in a more restricted sense, and was more focused on "agrarian structures" and their spatial expression at the level of the agrarian landscape. In fact, while the notion of "agrarian structure" had been used for a much longer period of time, it applied at the same time to the form, arrangement and layout of the fields, meadows, pastures and woods on the one hand, and to the size of the production units and the various associated tenure systems on the other: ownership, tenant farming, share-farming, etc.

By limiting the *system* to the *structure*, less emphasis was placed on the dynamics and progressive nature of agrarian societies, and on the systemic interaction already suggested by André Cholley or Claude and Georges Bertrand before that, giving the illusion that landscapes and agrarian systems were stagnating. The actual expression *agrarian system* was not a big success with geographers: no trace of "agrarian system" could be found in the voluminous *Vocabulaire de géographie agraire*

(Vocabulary of Agrarian Geography) by Paul Fénelon (1970). In the 2000 edition of the *Dictionnaire de la géographie* (Dictionary of Geography) by P. George and F. Verger, the word was still nowhere to be found, although the concept was almost evoked in the definition of "agrarian structure". Next to a static definition for which the agrarian structure was limited to habitat, field pattern and rural property, a much larger meaning was given: "agrarian structure would then be a '*combination*' of physical, biological and human elements in deep interaction". Therefore, the concept of system was an underlying concept, not yet that of the agrarian system. As to Roger Brunet in *Les mots de la géographie, dictionnaire critique* (The Words of Geography, A Critical Dictionary) (1993), under the "agrarian" paragraph, he wrote: "Agrarian system: traditional category of geography, at a time when 'system' had a limited meaning; described formally in particular the layout of space exploited by agriculture: relationship between the parties of the territory, parcelling and division, and sometimes elements of the agrarian system. Might have been extended to include the entire production method, the organisation of daily life, ideas and institutions". The first definition is clearly limited to the agrarian structure; the second is vaster, although particularly vague, since the author is manifestly not adhering to it. Finally, in the *Dictionnaire de la géographie et de l'espace des sociétés* (Dictionary of Geography and Societal Space), published in 2003,[1] the expression *agrarian system* appears in English as the potential translation of 'agrarian structure'... the confusion is understandable. The expression *agrarian system* is not even mentioned in *The Dictionary of Human Geography*, which is regularly edited by Johnston et al. (2000).

Nonetheless, the fact remains that comparative agriculture and agrarian geography have many things to share, a relatedness which geographers – being the first to speak of "agrarian system" – certainly have something to do with. Agronomists and geographers thus came together during a seminar organised by ORSTOM, entitled "Across Fields, Agronomists and Geographers", with the proceedings published in 1985 (ORSTOM 1985). Concerning this relatedness, in 2000 Paul Pélissier and Jean Pierre Raison spoke of "complicity between ruralist geographers and agronomists, which sometimes turned into symbiosis, a complicity which was – [they were] both convinced about it – an essential element for the fruitfulness of agrarian studies in the tropical world. After seeing this convergence developing, [they felt that] history would have to be written based on it. Individuals contributed to this history, and it is rather significant that the first agronomists open to the study of 'agrarian systems' included Deffontaines and Papy, both sons of geographers..." (Pélissier and Raison 2001: 14).

[1]Edited by Jacques Levy and Michel Lussault.

The "Agrarian System" of French Agricultural Economists

It was during the 1970s and the 1980s, and while it was felt that a more comprehensive level of analysis was needed to apprehend the ongoing agricultural transformations in Europe and the countries of the South, that a certain fad began, in France in particular, for the agrarian system concept. At the time, several agro-geographers or agricultural economists proposed their own definition, trying to apprehend the farm "environment" with a tool that was more comprehensive and made it possible to illustrate the multiple reciprocal interactions within this "environment", and between the environment and the farms: Deffontaines and Osty wrote, in 1977: "the working hypothesis is that there are areas in which the relationships among farms, and between the farm and the environment, show specific characteristics and are organised into systems we call *agrarian systems*" (Deffontaines and Osty 1977: 198). According to Vissac (1979), the agrarian system designates: "the association of productions and techniques implemented by a society with a view to satisfying its needs. It expresses in particular the interaction between a bio-ecological system, represented by the natural environment, and a socio-cultural system, via practices resulting in particular from the technical asset" (Vissac 1979).

The approach to agrarian systems was developed at the French Agricultural Research National Institute, Agrarians Systems and Development Department (INRA-SAD) in particular, and applied to geographical areas as different as the Vosges or Nepal (INRA 1977, 1986). This was also the era of the explosion of systems research in agriculture, combined with different levels of analysis, and of its non-equivalent *Farming Systems Research (FSR)* in the Anglo-Saxon world.[2] It led to the publication of important works at the beginning of the 1990s; a general idea of it can be found, for example, in *Systems Studies in Agriculture and Rural Development* (Brossier et al. 1993), or in the proceedings of the International Conference on *Recherches-système en agriculture et développement rural* (Systems Research in Agriculture and Rural Development) (Sébillotte 1996). The INRA publication of L. de Bonneval's work entitled *Systèmes agraires, systèmes de production – Vocabulaire franco-anglais*, (Agrarian Systems, Production Systems – French–English Vocabulary) (1993) also testifies to this interest. From then on, systems research was in vogue and particularly encouraged within the framework of the Association for Farming Systems Research and Extension (AFSR/E). Most works conducted by following this approach put forward the farm – not so much the higher levels – as the privileged level of systems analysis. Although they were conducted at the territorial unit level, these approaches often never went further than the systemic analysis of farms, never considering "the farm environment" as being itself systemic and addressing very little, or not sufficiently, the historical aspects, movementsand social relations. On the other hand, they were very much "finalised"

[2]On the compared analysis of the emergence and development of these two approach "families", FSR and systems research in agriculture, see Fresco's (1984) and Pillot's (1987, 1992) analyses.

by development objectives,[3] particularly when, at the end of a Rapid Rural Appraisal-type diagnostic, they led to the definition of technical recommendations (Khon Kaen University 1987).

It was on the contrary a search for coherence between the evolution of agricultural techniques on the one hand, and the economic and social system on the other, that led comparative agriculture to conceptualise the "agrarian system". To this end, it was while conducting research in the region of Monts Dômes, at the beginning of the 1970s, that Raphael Larrère insisted on the historical dimension and social organisation which, according to him, were underlying the very notion of agrarian system, and that, from then on, he moved away from the Farming Systems Research-type approaches in vogue at the time. He was defining an agrarian system as "an organised set of relations established historically between a determined social structure and the territory developed by it", insisting on the historical dimension and confirming the underlying principle that "as soon as they organised themselves socially, men maintained with nature relations that depended on the social organisation of production" (Larrère 1974).[4] He continued: "It is not the 'farmers' (a group of different individuals isolated from one another) who develop, and even less the 'future farmers' (a sub-group of farmers better equipped than others), but a social formation made up of different categories of farmers, as well as artisans and providers of various services, between whom a social division of work is established. A society endowed with its own operating laws and contradictions, and which can be the scene of conflicts between its components" (idem). The agrarian system concept found its vastest dimension when it was envisaged at the "social formation" level, an entity all the more encompassing since the modernisation of French agriculture, as well as the integration of many "developing" rural societies into trade, resulted in local idiosyncrasies being levelled out, as was believed at the time.

Head of the Department of Economics and Rural Sociology of the INRA from 1972 (before the INRA-based department of Agrarian System and Development (INRA-SAD) was created), then Professor at the Institut National Agronomique (INA) from 1974, Mazoyer endeavoured to define the agrarian system concept, by giving it a more dynamic and encompassing dimension than the definitions developed by Deffontaines and Osty (1977, *supra*) or Vissac (1979, *supra*), which remained limited to interactions between farms and the natural and economic environment, or those between "bio-ecological" and "socio-cultural" environments. The concept was then redefined as "a mode of exploitation of a given agro-system, historically constituted and long-lasting, adapted to the bioclimatic conditions of a given space, and meeting the requirements and social needs of the time" Mazoyer (1987: 11). M. Mazoyer specified: "the agrariansystem includes the following

---

[3]And were as such fairly subjected to "drifts" highlighted by Jean Pierre Olivier de Sardan (1994).

[4]On the approach in terms of agrarian system in the region of Dômes, see also Gilles Bazin et al. (1976).

essential variables: the cultivated environment and its historically acquired transformations, the production instruments and the labour force implementing them, the resulting environment *artificialisation* method, the social division of labour among farmers, handicrafts and industry and, consequently, the agricultural surplus and its distribution, trading relations, ownership relations and power relations, and lastly, all the ideas and institutions ensuring social reproduction" (idem, p. 12).

More recently, Mazoyer and Roudart redefined the agrarian system concept as "the theoretical expression of a historically constituted and geographically localized type of agriculture, composed of a characteristic cultivated ecosystem and a specific social production system. The latter makes the long-term exploitation of the fertility of the corresponding cultivated ecosystem possible" (Mazoyer and Roudart 2006: 51). It is precisely the reciprocal interactions between the elements relating on the one hand to "the cultivated ecosystem", and on the other to "the productive social system", which characterise the whole agrarian system.

To characterise further the complex content of this concept, I would argue that the agrarian system encompasses first of all the *mode of exploitation of a given environment*, i.e. one or several ecosystems, an exploitation method characterised by matching *technical qualifications* (tools, knowledge, practices, know-how), historically constituted forms of environment artificialisation, specific relations between the different parts of the ecosystem(s) used, and one or several mechanisms for reproducing the fertility of the cultivated lands. It also includes *the social relations of production and trade* which contributed to its establishment and development (*conditions of access to resources* in particular), as well as the conditions for the *distribution* of the resulting *value added*. It further includes a limited number of *production systems*, *differentiation mechanisms* between these systems and *their respective trajectories*. Finally, it includes the characteristics of *specialisation and social division of labour* within the industries, as well as *economic, social and political conditions* – the *relative price system* in particular – which determine the conditions and consequences of the integration of producers into the international market.

Therefore it is the understanding of the agrarian system's *functioning* which is at the centre of the approach to comparative agriculture, the agrarian system concept being "an intellectual tool that leads to apprehending the complexity of any form of true agriculture, by methodically analysing its organisation and functioning. This concept also leads to classifying the countless forms of agricultures identifiable in the past or observable today in a limited number of systems, each one being characterised by a specific kind of organisation and functioning" (Mazoyer and Roudart 1997b). This approach was at the root of the theory of agrarian systems taught from the very beginning of the 1980s by Mazoyer, at the Institut National Agronomique of Paris-Grignon, before being finally published in his book *Histoire des agricultures du Monde, du Néolithique à la crise contemporaine* (1997a), later published in English: A History of World Agricultures, From the Neolithic Age to the Current Crisis (2006) (see Box 3.1).

Box 3.1: Marcel Mazoyer and the Theory of Historical Transformations and Geographic Differentiation of Agrarian Systems

The many forms of agriculture in the World can be analysed and classified into a few types, each one characterised by a specific organisation and functioning method, and a specific process (agrarian system). "The theory of the evolution of agrarian systems is the tool that makes it possible to represent the continual transformations of agriculture in a region of the world as a succession of distinct systems, forming a definite historical series[. . .], the tool that makes it possible to apprehend and explain in broad outlines the geographical diversity of agriculture in a given epoch" (Mazoyer and Roudart 2006: 51).

> . . . by methodically explaining the structure and functioning of an agrarian system, a sort of archetype is formed that necessarily provides a coherent and harmonious image to the corresponding species of agriculture. This archetype, which clarifies the rationality of a particular species of agriculture in space and time, that is, its *reasons for being*, expanding and surviving through adaptation, is necessary in order to identify and classify the observable forms of agriculture belonging to this species and recognise their particularities and possible failures. (idem, p. 52)

The agrarian system concept is then used to classify and characterise the agriculture of very vast geographic wholes. It actually concerns most often a *family* of agrarian systems which are mainly designated and characterised as follows:

- Forest Agrarian Systems: These are characterised by a farming method dominated by slash-and-burn agriculture in wooded environments, involving a few years of cultivation alternating with long periods of regrowth of shrubs then woods. In the quasi-total absence of deep ploughing tools, only a land that has been let lie fallow with trees over a long period of time can eliminate the herbaceous layer which is competing with the cultivated plants. Moreover, livestock farming being also barely developed in these types of agriculture, where the forest environment is not suitable for it, the regrowth of trees is also what leads to restoring the fertility potential of the environment. Often associated with methods for accessing the land regulated at the level of the village or the lineage, it was possible for these systems to be sustainable and preserve the forest ecosystems exploited, as long as population density remained low and large areas of the territory were not dedicated to other uses (perennial planting for example).
- Hydraulic Agrarian Systems: The progressive series of the hydraulic systems of the Nile River Valley, which is archetypal of these types of agriculture, illustrates the different stages of the increased artificialisation of the environment, from the first flood-recession basins built for winter crops from the 6th millennium onwards, to the erection of the largest works

(continued)

Box 3.1 (continued)

making irrigated crop possible today in any season; via the progressive expansion and unification of the flood-recession basins, their supply in flood water via large canals and their protection thanks to dykes; then via the progressive development of irrigation during Antiquity (Archimedes' screw and bucket wheel), during the Middle Ages (animal-drawn water pumping), then in the modern era with the first major diversion weirs in the nineteenth century. The issue of building, managing and maintaining these hydraulic infrastructures by a centralised public authority, has always been central to the functioning and sustainability of these systems.

- Inca Agrarian System: This is a mountain agrarian system made up of complementary terrace subsystems: irrigated maize, bean and cotton crops in the oases of the coastal desert and the lower sides on the Pacific coast; rainfed and irrigated maize, bean, lupine and quinoa crops in the valleys of the Andes; potatoes, Andean tubers and Camelidae breeding at high altitude; maize, bean and manioc crops in slash-and-burn agriculture on the Amazonian side of the Andes cordillera. It is the unification of these different terrace agrarian systems, thanks to a system of trading relations, centralised administration and major infrastructures sustained by levying a tribute in the form of labour, which gives this agrarian system its originality.
- Agrarian Systems Based on Fallowing and Animal-Drawn Cultivation with the Ard in the Temperate Regions of Europe: Born, as early as Antiquity, of the association between pastoral farming and rain-fed agricultures in the already largely deforested Mediterranean areas, and in a large part of Europe with a temperate and cold climate, these systems are characterised by differentiating between (1) the *ager*, a set of fields worked on with an ard and as a two-course fallowed rotation; (2) the off rotation *saltus* reserved for herd grazing, and with the possibility of fertilising the cultivated lands by parking the herds at night on the fallow lands; and (3) the *silva*, which supplied wood and game. Although this system is characterised by cereal yields of a few quintals only per hectare, and by livestock farming still largely dependent on the lowest level of winter forage, its emergence appears as an adequate response to the crisis of the formerly predominating forest systems.
- Agrarian Systems Based on Fallowing and Animal-Drawn Cultivation with the Plow in Cold Temperate Regions: These systems are characterised by the significant development of winter forage supplies, thanks to the expansion of meadows where grazing animals are prohibited, and at the same time by the development of transport means (carts) and the storage of forage (hay-barns); by the maintenance also of larger herds, their confinement in cowsheds during the cold season and the production of real

(continued)

Box 3.1 (continued)

manure (litter), which is carried to the crop fields thanks to the development of animal-drawn transport (tipcart); by the concomitant increase in cereal yields, the generalisation of the three-field system, particularly in Oceanic and Northern Europe where only one third of rotated areas are reserved for fallowing, and by the use of much more efficient ploughing tools (ploughs).

- Agrarian Systems in Temperate Regions with No Fallowing Stemming from the First Modern Agricultural Revolution in the eighteenth and nineteenth Centuries: Former acreage lying fallow, occupied from then on by fodder root and tuber crops as well as forage crops, and offering food production as well as enough forage to maintain a larger herd in winter. The result is an increased production of manure and therefore a substantial improvement in cereal yields. But preparing fallow lands for cultivation supposed several social transformations beforehand such as, among others, putting an end to compulsory rotations and to the right of commonage prevailing until then in Western Europe.
- Motomechanised, Specialised and Chemicalised Agrarian Systems Stemming from the Second Modern Agricultural Revolution: Equipped with increasingly powerful machinery, these production systems are from then on specialised in the production of a small number of foodstuff, stemming from selected vegetal and animal materials, with cultivating and breeding conditions being increasingly artificialised, and with the systematic use of artificial fertilisers, pesticides and veterinary products. While farmers only represent a small portion of the working population, and while their labour productivity has increased tenfold, the horizontal division of labour (specialisation) is doubled by a vertical specialisation of labour, and the upstream and downstream industries take on an increasing portion of the activities formerly taken care of, at least in part, by the actual farmers.

The actual system concept gives rise to the notion of balance and reproducibility or, in today's words, sustainability. The concept of *fertility*[5] and the study of its reproduction mechanisms turn out to be important, whether reproduction is tackled at the level of the cropping (*infra*), production (*infra*) or agrarian system. Since the works of Claude Reboul in the 1970s and the 1980s (Reboul 1977, 1989), it is particularly clear that fertility is something other than a set of "natural" conditions, that it is just as much the result of economic and social processes and the product

[5]Sébillotte (1989) insisted understandably on the inherent dangers of making inconsiderate use of this word, to in fact designate different realities or even simple and mostly anthropomorphic representations. He prefers to speak of *land cropping suitability*, with the different components of fertility – envisaged within the conceptual framework of the cropping system – represented either through yield, cultivation costs, requirements for the application of cropping techniques or the risk inherent to their application (Sébillotte 1989).

of a history, as the result of an "agronomic" evolution stricto sensu.[6] That is why thoughts on the conditions underlying the reproduction of fertility and, more generally still, on the biomass levels of exploited ecosystems, their evolutions as well as the potential transfers of biomass between the different constituents of these ecosystems, are non-conveyable in any analysis in terms of agrarian system.

Generally speaking, it is the maintenance and reproduction mechanisms of the *conditions* underlying the exploitation of these ecosystems, i.e. those underlying in particular the maintenance of fertility and balance of the exploited ecosystems, but also the requirements for the reproduction of the material and human means of their exploitation, as well as the stability of the prevailing social relations, in short everything that partakes of what we could call *a regulation mode*, which is an integral part of the agrarian system and partakes of its definition. "It is this global coherence of reciprocal determination and reproduction of the different elements of the system that, in fact, establishes the unit of the agrarian system concept. Furthermore, as important as it is, analysing interrelations between the different levels of the system (...) is not enough to found the notion of system. What matters is to bring out the fundamental logic of the system's reproduction, so as to characterise its unity and outline" (Kroll 1992: 12).

For all that, this coherence does not mean absence of internal contradiction, differentiation or conflict. On the contrary, the accumulation systems and the differentiation mechanisms, those of the production systems (*infra*) in particular, characterise the system itself; differentiation *makes* the system. Sticking to the "weak" description, i.e. defining the system on the basis of the interdependent relationship that exists between the units or elements of that system, is not enough. "What matters is that the system defined in its generality (or its purity) entails 'dynamisms' on the very grounds of the differences making it up".[7]

3.2 Scales and Boundaries

From Agricultural District to Country, What Scale Should Be Privileged?

Using the agrarian system concept is not an easy thing, particularly in certain situations. There is for example the problem of the limits or boundaries to be attributed to an agrarian system, and therefore that of the more specific definition

---

[6]Symmetrically, we can show that erosion, decrease in soil fertility and ecosystem degradation in the general sense, can sometimes be construed as being the result of economic and social processes more than that of "natural" mechanisms. Fertility, like the capital with which it is sometimes confused, partakes of a differential accumulation and – that's the other side of the coin – of a no less differential erosion (Cochet 2001).

[7]Balandier, concerning social dynamics and systems (1971: 49).

of space, where its application would be more pertinent. Are we talking about the village, the "agricultural district", the region or the country[8]?

The village or "rural community" constitutes a first scale of observation and analysis, where the relations existing between the basic production units, often reflect a specific method for farming ecosystems, leave their mark on the landscape to the point of being "read" in it and form a historically constituted, socially determined and long-lasting coherent whole. Many examples could be mentioned where the agrarian system concept was in fact used on that scale: in Western Africa with the village agrosystems of Jouve and Tallec (1994) or the *terroir* studies conducted under the supervision of geographers Sautter and Pélissier; in the Andes cordillera with the community organising collective cropping plans in altitude and the management of territories (Morlon 1992). In the old European agriculture (before the post-WWII transformations), many villages could have been analysed in this way. As such, it is very often on the scale of the territory that the agrarian landscape best reflects the spatial expression – *i.e. what can be seen* – of the agrarian system.

A large number of villages can also leave the same mark on the landscape, with the latter presenting common characteristics and reflecting common regulations on a much vaster space. Moreover, what is being played at the level of the territory depends also on elements situated outside of it and cannot therefore be fully understood at that level. That is why this scale of analyse is too limited for an overall comprehension of agriculture, and it seems to me that the agrarian system concept calls for a much vaster application area. All villages and/or communities would then relate to the same agrarian system. The activities of these villages and/or communities would leave a similar mark on the landscape and be organised around the same rules and institutions. The geographical limits of the agrarian system would then be determined by the territorial extension of these rules and common practices (Jouve 1988). In this case, the village agrarian systems only constitute an intermediary – although indispensable – level of understanding of the agrarian systems, a level included between the production system level (the farm, *infra*) and agrarian system level (the region).[9]

The agrarian system concept can also be used to classify and characterise the agriculture of much vaster geographical wholes, as proposed by Mazoyer in his theory of agrarian systems *(supra)*.In it, he distinguishes forest agrarian systems,

---

[8]L. de Bonneval's French-English dictionary, already mentioned, testifies to the vagueness maintained on this issue (op. cit., pp. 172 and 175).

[9]With Thierry Link, Eric Léonard and Jean Damien de Surgy, in Mexico, we had introduced, at that level of analysis, the concept of *sistema social de producción* (Cochet et al. 1988). The approaches developed by certain projects and research departments intervening in Africa (e.g. IRAM) as regards "*terroir* management", also privileged this level of analysis and intervention.

the hydraulic agrarian systems of the Nile River Valley, agrarian systems based on fallowing and animal-drawn cultivation with the ard in temperate regions etc., where the use of the plural actually indicates that it often concerns a *family* of agrarian systems (see Box 3.1).

But the identification and limits of agrarian systems are still an issue, as pointed out by Jouve (1988). Is it in fact absolutely necessary to resolve this issue? And would the space in question be geographical or would it be just as social and political? Isn't it left to the researcher to determine, in a flexible and pragmatic manner, the most suitable space for this systemic construction or, rather, the adaptation of the concept in order to better apprehend the space under study and that which "fits" better the reality under study? Nothing prevent us actually from conjoining scales and combining approaches to distinguish, on the one hand, a "local" agrarian system that would be relevant at the level of a small region with a "homogeneous problematic",[10] and on the other an encompassing agrarian system resulting from the aggregation or grouping of several and actually very largely interdependent local agrarian systems.

This is also the approach proposed by Marielle Pépin-Lehalleur and Gilles Sautter in Mexico: "The idea of basic agrarian systems fundamentally brings into play in each case the association of a number of components, which are not simply juxtaposed but more or less completely interdependent. It supposes on the other hand (...) a degree of spatial cohesion between the functional units – villages or communities, farm enterprises or family enterprises – partaking of a similar agrarian system" (1988: 22). In the mind of these authors, the regional agrarian system is "a higher level of systemic interaction that would somehow come to oversee the different basic agrarian systems, more or less clearly located" (idem).

The African highlands of Rwanda and Burundi offer another example, where the agrarian system concept can be efficient and pertinent in understanding and characterising agriculture in that part of the world, which is dealing with the same "agrarian system"; which does not prevent us in any way from distinguishing, on a more detailed scale, different agrarian systems sometimes even fairly contrasted, despite certain shared rules and characteristics (Cochet 2001).

Moreover, with comparative agriculture, it is possible on the one hand and for cognitive purposes, to classify and arrange diversity into a restricted number of "types", i.e. agrarian systems as proposed by Mazoyer's theory of agrarian systems (see Box 3.1); and on the other hand, not to restrict this diversity too much, particularly when the researcher, involved in a finalised research, must produce results to identify, formulate and implement within reasonable time limits, development projects or programmes adapted to each situation.

[10]What Jouve defines as "basic agrarian system", (1988, op. cit.: 12). This is also Deffontaines' preferred scale of analysis: "The local agrarian system: the level of analysis of land use and development logics, a level where the technical, economic and ecological balance sheets of the agricultural activity can be carried out, a level where conflicts and solidarities linked to the practice of this agricultural activity can be grasped" (1991: 32).

Spatial Bursting, Temporal Telescoping

Another difficulty comes up when certain types of agriculture appear increasingly difficult to delimit in space, for example when an important portion of the workforce is involved in long distance seasonal or even long term migrations. The case of migrants originating from the valley of the Senegal River who settled in the East of Paris, that of Mexican farming families of which one or several members have, for many decades already, been coming and going or have settled in California, or that, more recent, of vast equatorial regions largely emptied of their lifeblood in favour of Spanish irrigated agriculture, illustrate the situation. But the fact that an important portion of the workforce is absent part of the year, the fact that a third, half or more of household revenues come from transfers originating from foreign countries – proof that agrarian systems could not function in isolation but still remain open systems – does not question the pertinence of the approach in terms of agrarian systems, in that the elements mentioned above often reveal the crisis while also strongly contributing to its reconstitution in another form.

A certain "acceleration" of history during the last 50 years (contemporary agricultural revolution in the North, the brutal integration into trading of many agrarian societies in the countries of the South, among others) has also made the use of the agrarian system concept more delicate. Indeed, it is easier to analyse a relatively "stable" situation, and as such *build* the agrarian system, i.e. the systemic representation leading to a global apprehension of agriculture, than to engage in the same exercise when everything moves so fast that the different elements of the system, as well as their barely established reciprocal interactions, transform again. Whether we are talking about the French countryside, with the establishment of increasingly effective production systems and the massive restructuring that comes with it and makes it possible, or whether we are talking about the brutal opening of the farming communities of the South to competition and its consequences, it is often more easy to recreate the "old" agrarian system and take apart the mechanisms of its deconstruction and transformation, than to characterise the current or future agrarian system in clear terms. Would it be easier to access and mark out the agrarian system concept on the scale of long term history, than on that of the accelerated transformations of contemporary agriculture?

Similar difficulties crop up when agrarian dynamics on the frontier need to be understood and anticipated. One cannot study the dynamics of a colonisation front as if it was an agrarian system clearly marked out in space (within geographical limits, beyond which one would swing over to another agrarian system), and in time (a timespan, wedged in a periodisation where we would clearly see different agrarian systems succeed one another). People, their technical background and their practices move at the rate at which the colonisation front advances, as these practices transform, evolve and adapt to the changing conditions of the "natural" as well as human, economic and social environment. As such, it is not about an expansion phase (in space) *identical* to an agrarian system, because agricultural practices, as well as social relations, evolve *behind* the front under the pressure of a growing demographic density and land division, and in comparison with the perspectives of evolution and movement which some always benefit from, as long as the front remains active. It is not either about a strictly temporal process where one would see different agrarian systems succeed one another on a given territory (a space), because the progressive processes at a given place are also largely determined by what is happening behind and beyond the front.

These situations, which have also been very frequent in the last 50 years, call for a certain adaptation of the tools or concepts used. In this case, it is the *process* that needs to be analysed in all its spatio-temporal components and understood. The problematics of going from a forest agrarian system on the margins of the Amazon Basin, in Bolivia, Peru or Ecuador, to one or several post-forest agrarian systems involving significant migration movements in particular, can only be analysed within this framework. In a completely different context, agrarian systems of "ex-forest" Ivory Coast cannot be apprehended without putting them back into the wider context of the cocoa/coffee front, which has been sweeping over half of the southern part of the country in the last 50 years. Each area, each period, can only be understood as part of a whole, a process, and it is this internal process which in the end makes it possible to characterise the agrarian system *in motion*.

Box 3.2: Agrarian System in Motion: The Frontiers of Extensive Animal Production in Mexico with, in the Background, Share Livestock Farming (According to Cochet et al. 2010a)

The low altitude wet tropical zones of Mexico, on the Atlantic side in particular which receive more rains and favour grass growth, have experienced the spectacular development of cattle farming for meat production during the second half of the twentieth century, the dominating activity of a real pioneering front to the detriment of the vast forested areas that prevailed before that. Although this movement unfolded over almost half a century and concerned vast regions, it is characterised by the remarkable continuity of the exploitation mode of the extensive environment, based on grass fattening and limiting investments to very little, on the one hand, and of the related social practices, based on share livestock farming or livestock lease and ensuring its socio-spatial distribution, on the other.

This general movement concerning the development of cattle farming for meat production, took place at first from the 1940s to the 1960s in all the dry and humid regions of the tropics, that were still little concerned by the agrarian reform: during this entire period, when cattle farming for meat production was synonymous with large farms, the parallel expansion of share-crop contracts was still involving, as was the case in pre-revolutionary Mexico, the powerful land owner with a monopoly on the one hand, and the precarious farming community, often made up of migrants and devoid of production means, on

(continued)

Box 3.2 (continued)
the other. The workforce present on the cattle farms was often made up of land-clearing farmers, tasked with increasing fodder areas. In exchange for being granted the right to produce one or two cycles of food crops (maize and beans) with slash-and-burn agriculture, they were to take care of the sowing of forage grass on the same plot, with a view to giving the grassland back to the land owner a few months later, most often together with the crop residues (maize canes and leaves). Despite the diversity of possible arrangements, tenant farmers who were maize producers became grass sowers at the service of powerful land owners. They had to relocate their crop fields and sometimes their houses as the pastures progressed on the front, in a South–East direction.

In the 1960s and the 1970s, the legislative tools of agrarian reform (land endowment process) were increasingly used for colonisation purposes. Around 32 million hectares were thus distributed in the 1960s alone to more than 500,000 beneficiaries, many of which in the form of *ejidos* plots of 20 ha or so in the tropical forest. The comparative advantage of wet tropical zones as regards forage production, as well as the favourable evolution of the relative price of beef, particularly in relation to maize and its by-products (pork), made of the farming of grass-fed cattle for meat production an attractive activity, including for the newly endowed small tenant farmers.

That is why a real specialisation in cattle farming for meat production emerged in most regions of the Mexican tropics, particularly among agrarian reform beneficiaries. Yet the modest size of their farm could have justified, in another political context, the implementation of more labour intensive production systems, and shown more value added per hectare (perennial plantations, for example).

While the development of land endowment began necessarily with the establishment of food crops (maize and beans), via slash-and-burn agriculture at the rate of 1 or 2 ha per year (a surface area which was limited by the hand tools at the disposal of the migrants, and by the difficulty of selling a potential surplus), the deforested surface areas were progressively sowed with forage grass then enclosed, thereby progressively contributing to the improvement of the ground.

Most beneficiaries of the colonisation programme, who were too poorly endowed in production means and had no access to credit, could not begin a breeding unit for lack of capital. They had no other choice but to turn to the big livestock farmers who came before them, to obtain the few cattle heads that were essential for starting their new activity, against the payment of half the product to the owners of the cattle. This new social relation gave birth to livestock lease or share livestock farming.

Agrarian reform had just created the conditions for the spectacular reversal of this social relation (reverse tenancy). Relaunched on a massive scale

(continued)

Box 3.2 (continued)
for the purpose of agricultural colonisation, the reform enabled hundreds of thousands of small farmers to become holders of a land right that was perennial and secured in a frontier situation. But failing to extend access to production factors, fixed assets and working capitals in particular, the agrarian reform led to the profusion of agrarian contracts where the land holder – the lessor – saw his role being limited to the production of forage, and found himself in a situation of dependency vis-à-vis the "lessee" who was then able to supply the capital required for the production process.

These contracts included, for example, the supply of barbed wire and bull-calves of 16–18 months old (i.e. around 200 kg of live weight), at the rate of one head per hectare and for a fattening cycle of 18–24 months. They included possibly the veterinary products that were essential for fattening activities, as well as, more rarely, certain cash advances. Once the animals had reached a live weight of 450–500 kg, the sponsor decided when they were to be sold, depending on the market situation and on his own economic constraints. The earnings corresponding to the weight taken on by the animals were then shared between the two parties, once all input and cash advances of the cattle owner were deducted. The share-farmer was responsible for all the pasture production and maintenance costs (cutting ungrazed grass and repairing fences), as well as potential animal losses (death rate). What is more, in case a contract was not renewed, the sponsor would take back the fixed assets (barbed wire) invested by him on the farm of his share-farmer.

Livestock leases prevailed on the Atlantic side, with the store cattle supplied from the dryer tropical regions of the Pacific side. Generally, they corresponded to the opening pioneering phase of new productive areas. But with time, breeder contracts were also proposed to small tenant farmers in the few regions of the South-East of Mexico which were less conducive to grass fattening, in the "stabilised" regions behind the frontier, or even, much later, on the actual agricultural colonisation frontier.

Small producers were then appealed to by stock farmers to ensure their supply in store cattle, while the latter were specialising in cattle fattening. This time, the corresponding arrangements included handing over heifers as well as one breeding bull, and provided for the distribution of new-born animals between the two parties, with the stock farmer taking the males as a priority and leaving the females to the share-farmer. This type of arrangement was sought after by small forage producers, since it gave them access to breeding cows that, in the long term, could ensure reproductive autonomy faster than in the case of share fatstock farming.

The fact that, for the share-farmer, the process of building up a herd was slow, guaranteed major livestock operators a certain stability of the land on which they could develop their activities, since the gradual saturation of

(continued)

Box 3.2 (continued)
pastoral land was compensated by the progression of the agrarian frontier, and the settlement of new generations of farmers. While behind the frontier, the accumulation process enabled the beneficiaries of the agrarian reform, over a period of time equal to at least the formation of a new generation, to become autonomous after building up a small herd of 15–20 breeding cows, the advance of the agrarian frontier opened up new lands to the major stock farmers, lands on which they could expand their clientelist networks as needed. Thus, the movement of the colonisation front, during 40 years, laid the foundations for the similar renewal of the technical systems and social relations, which the prosperity of the major private stock farmers relied on.

Share livestock farming was not only at the centre of the accumulation system characteristic of the agricultural frontier; it was also its main mode of regulation. Through the generalisation of livestock leases, the relationship to land ownership became a secondary element of negotiation, just as the position of the leaseholder weakened, while controlling the capital secured the hegemony of the livestock owner and, in fact, the decision centre of the livestock farming system established. This new "tenant farmer", potentially deprived of any property ownership, but able to supply almost all the capital (fixed assets, working capital, share capital etc.) required for the production system to function, takes over from the powerful land owners of yesteryear in the social and power relations at the local and regional levels.

In fact, agrarian systems are *always* on the move. In order to apprehend and understand their *structure* and *functioning*, one needs to define a "state" of the system at a given moment of its history, i.e. to "imagine it" as being stable, long enough to catch a glimpse of its basic interactions and mechanisms, a non-conveyable step to perceive and interpret the *movement*, to discern the conditions of its durability or, on the contrary, the causes of its imminent crisis.

The Agrarian System Goes Beyond and Encompasses the Sphere of Primary Production

It is possible that the growing specialisation of farmers in the primary production segment alone, the growing importance of upstream and downstream industries in the supply of means of production and processing of products, the distance created between producers and consumers, urbanisation, as well as national and international migrations, made some see the agrarian system concept as being less effective because societies (even *rural* ones) were decreasingly "agro-agricultural".

In Western Europe, the contemporary agricultural revolution was characterised, in particular, by the outsourcing of many tasks formerly carried out by the farmers themselves: activities involving the self-supply of production means, the first processing at the farm, as well as the transport and marketing of products. Farmers' specialisation and the fact that they were confined to the production of raw materials, and were forced into abandoning any activity for the production of tools and inputs, went hand in hand with the development of upstream industries (mechanisation and various equipment, chemical industry, bio-technology etc.), downstream industries (the food-processing industry, packaging etc.) and the transport sector.

This evolution led the agricultural sector to experience a drastic reduction in a number of functions and tasks formerly carried out by farmers or their immediate artisan neighbours. But the terms of the division of labour throughout the food production process, from the manufacture of the production means used by farmers, right down to the food landing in consumers' plates, partake of the characterisation of the "agrarian system". This new division of labour, and the new industries stemming from it, are all *constituents* of agrarian systems born of the agricultural revolution. Consequently, the application of this concept cannot under any circumstance be limited to the sphere of non-processed primary production.

The economy of the food-processing industries, the networking approach, experienced a spectacular development in the last decades, and widely showed the extent to which production could be "run by the market".[11] Today, producers from around the world are condemned to have their eyes riveted on "quality standards" (often corresponding to the standardisation of products and regularity of their flows, but also to health and phytosanitary standards, criteria of quality or origin, etc.) if they hope to place their production on the world market under appropriate conditions. While the production of each farmer or each region of the world is *partly* run by the market and often increasingly "formalised by contract", it is also put out to *competition* on an increasingly standardised market. The conditions of this opening for competition, and the *real* comparative advantages which one production or another can enjoy in a given region or country, ought to be studied carefully. The measure of this comparative advantage must take into account opportunity costs, attributed by the producer and his family to the different means of production at play, and considering the minimum revenue bearable under the socioeconomic conditions of the time and place.

While giving the greatest significance to these determinants, and taking the greatest care to locate its genesis and development, the economic approach in comparative agriculture remains nevertheless centred on the actual production process, and within a framework of analysis meant to be systemic, particularly at the production system level (Chap. 10). Indeed, at that level, and for the great majority of agricultural products exchanged in the world, networks remain intermingled and interdependent at their peak or point of departure, i.e. at the level of the production node, simply because, in the end, a farmer being completely specialised in only one

[11]Concerning the networking approach and its recent developments, we refer the reader to the excellent review on the cotton industry in Africa written by Philippe Hugon, within the framework of a conference organised in 1992 on "Institutional Economy and Agriculture" (Hugon 1992).

type of production is a fairly rare occurrence. The time and means of production devoted to export crops (e.g. coffee, cocoa, groundnut and cotton) depend closely on the opportunity cost of these very factors, when they are dedicated to each one of these crops and/or food crops. The assessment of the respective costs which, for the major part, control whether a farmer will emphasise one type of production or another, can only be carried out at that level of analysis and in systemic terms. The same could be said for many of the European producers who are rarely involved in exclusive monoproduction or, when they are, who are anxious to get out of it.

More generally, it is clear that the unprecedented increase in long distance trading, with contemporary globalisation being nothing more than its extension and completion, makes agrarian systems more open than ever. Some of the conditions of their reproduction sometimes need to be sought far away from their geographic area of expression. Jean Christophe Kroll described this situation very well: "As soon as non-agricultural spheres of activities become dominant, as soon as their prevailing production and distribution ratios tend to structure society as a whole, the reproduction logic of the conditions of the agricultural activity relates to determinations that are increasingly external to agrarian systems per se. Consequently, long distance markets play an always more important mediation role between the production and consumption of foodstuffs, so much so that today there is no longer any immediately perceptible coherence between the evolution process of the production capacities of agrarian systems across the world, and the needs to be met in food requirements. In the same way, the fact that agriculture is increasingly using means of production bought outside the sector, blurs any coherence directly perceptible in production places between conditions for the reproduction of social relations of production, such as cultivated ecosystems, because the many mediations operated by the markets sometimes appear so complex and confused" (Kroll, op. cit.: 13).

With a view to integrating all the businesses involved in food production into a similar analysis (also conducted in terms of system), the expression "food system" (or agro-food system) was used to describe the whole process of production (including agro-supply), transformation, putting the products on the market as well as food distribution, therefore including de facto related consumption patterns. In the French case, this whole included close to 1.2 million businesses in 1995, farmers included (Rastoin 1996), a sector of activities that was studied like we could have studied the "motor industry" or any other sector. But the standardisation of process and markets on a worldwide scale, led the same author to speak of "tertiarised agro-industrial food model" (Rastoin 2008). Run by the markets and largely dominated by a restricted number of very large firms intervening upstream in particular (farm equipment, seeds, pesticides and fertilisers), and in the mass marketing sector, the tertiarised agro-industrial food model would somehow oversee all agrarian systems worldwide, by bringing in the largest portion of raw materials – the deposit – produced at their level. This concept – and scale of analysis – is useful in ensuring the coherence of the mechanisms at play throughout the agri-food sequence, and in emphasising the growing interdependence of these different elements. For all that, the concrete analysis of actual agricultural production processes and their territorial anchorage, on a necessarily more restricted scale, make it possible to identify the conditions and mechanisms of the generalised competitiveness of producers, and at the same time to measure the impact of the world food system on these very producers and their territories. Furthermore, with the recent search for so-called "local" networks, the development of processing units aiming at creating more value added locally, or the reinforcement of networks that could have played a role in the structuring of territories (certain AOCs in France in particular), one would like to see the emergence (or perhaps only endurance) of more autonomous agri-food systems vis-à-vis international markets.[12] The ambiguous expression "territorialised network", which is introduced sometimes in the discussions on territorial development in France,[13] also testifies to this desire.

3.3 Production System, Cropping System and Livestock Farming System

The study in terms of agrarian system also calls for the use of concepts for which efficiency and pertinence are measured on other scales of analysis, particularly those of the production unit on the one hand, and the cultivated plot or herd, on the other.

The Concept of Production System

This concept made a strong appearance in the field of agricultural economy during the 1970s and the 1980s. Rather than doing a – necessarily fastidious – review of the different definitions proposed for the term "production system" as applied to agriculture, or its Anglo-Saxon equivalent *farming systems*,[14] we will content ourselves here with recalling what – to me – appears essential in this concept, before insisting on its usage in comparative agriculture.

First of all, the scale of analysis which the application of the production system concept relates to, is generally that of "the farm" or "basic production unit". In many regions of the World, this unit is most often centred on the farmer's family, but the production system concept could just as well be implemented in the case of much vaster farming business-type production units, although certain difficulties can crop up, as will be seen further on. As suspected, this scale of analysis is quite primordial, since it is indeed the production units, the farms, which constitute the basic

[12] In this regard, see the "localised agri-food system" concept and the conference organised on this issue in 2002, on the initiative of the UMR Innovation of INRA, CIRAD, CNEARC and the Universities of Montpellier I and Versailles-St Quentin.

[13] For example, at the time when the Ministry of Agriculture and Fisheries proposed "land farming contracts" (CTE) to farmers (for the 2000–2005 period).

[14] In this regard, see Fresco (1984, op. cit.), Brossier (1987) and Pillot (1987, op. cit.).

components of the rural fabric, the level of organisation of the productive process in agriculture, that where production networks cross and intermingle; the basic components between which neighbourhood relations, solidarities, contradictions, conflicts and differentiation mechanisms are formed. A level of analysis which is all the more important since it is often at that level, i.e. through surveys conducted with farmers, that the researcher's first "contact" with the field is established.

Let us recall that while a production system is indeed made up of a set of elements interacting reciprocally, three simple ideas are necessarily associated with it: (1) this "set" functions, its parts allow it to work or, in this instance, to produce; (2) there is the idea of internal coherence, where elements are not juxtaposed in a haphazard way; (3) a system is endowed with a certain stability or "reproducibility", or the fact that its condition is being changed transforms it into another system, which is itself more or less sustainable compared to the system it succeeds.

The publication, in 1978, of P. L. Osty's article on "Farm Holding Seen as a System", was unquestionably a turning point in the French approach to farm holding, and was followed by many works where this new way of looking at things led to better understanding the farmer's "practices" and "choices", and thus to adapt agricultural advice. As such, the farm of Mr. X was analysed comprehensively and by seeking to bring out the many relations existing between the different units developed on the farm, the family structure, the available plots and the machines at his disposal etc. It was studied as if it formed a system. Such an analysis made the formulation of personalised advice possible. Applying the production system concept to a farm (rather than a group of farms) with a view to offering individual advice had already been conceived and developed by Chombart de Lauwe and Poitevin in a managerial perspective (1957).

For all that, and although the concept can indeed be "applied" to a specific farm and contribute to understanding how it functions, the individual application of this concept brings it to quickly reach its limits. Such an approach can only take shape, as an analysis in terms of production system, when cleverly selected interviews and farm visits are repeated long enough to make the elaboration of a real *typology* of farms possible, each type then being represented and explained through one – and only one – *production system*. Reality being what it is, the need to classify in order to better understand, necessitates that farms be grouped into types. Because farms belonging to a specific type are fairly similar in structure, practice as well as functioning, it will be possible to analyse and characterise them through *one* production system, which then becomes the modelled representation of the farm type under consideration. The concept will therefore be applied to one group of farms with the same range of resources (same range of surface areas, same level of equipment, same number of labourers), farms that have comparable socioeconomic conditions and practice a similar combination of productions; in short, a group of farms that can be represented by the same *model* (Cochet and Devienne 2006).

A farm, as such, ceases to be an object of modelling and does not constitute a production system, no more than a production system can explain how a farm functions or why a farmer makes a particular choice. That is why this approach cannot and must not, in our opinion, lead to individual and personalised advice.

In this perspective, a production system is therefore an intellectual construction, i.e. a *model*, which is useful in trying to understand the origin, functioning and future perspectives of a specific farm *type*, within a given agrarian system. The issue consists here in understanding the dynamics of an agrarian system as well as the evolution of the *production systems* making it up.

In their quest for production systems, economists and agricultural economists were confronted with major difficulties in locating the identity and boundaries of production units within the framework of rural societies in Sub-Saharan Africa, since it was difficult to include the production process into residential, accumulation and consumption units (Gastellu 1979; Colin and Losch 1994; Chia et al. 2006). More recently, the contemporary development of new institutional forms of agriculture, from which labour and capital are increasingly dissociated, creates the same type of problem for whoever seeks to apprehend the production process. Where are the respective centres of gravity positioned, and where can the boundaries of the different production units be located in the case of agricultural production companies intervening among many small owner-occupiers, who are decreasingly becoming farmers and increasingly owner-labourers?

Identifying the outlines of the production unit and therefore the perimeter that gives meaning to the concept of production system, sometimes meets with a few difficulties, as does the identification of the boundaries of an agrarian system with geographically very remote elements (*supra*). This in turn hinders the study of the actual production process and the use of the concept. However, irrespective of this former or recent disconnection between family and farming units, and irrespective of the various origins of the workforce, the land and the capital (with the centre of gravity of the process relying decreasingly on the family), the fact remains that the meeting of these production factors with a view to enabling a production process, still constitutes a system (or at least it is a possibility we need to suggest in order to understand these processes).

Technical Choices Embedded in Social Relations

The differentiation of production systems within the same region and, beyond that, the unequal development of regions or countries, result for a very large part from unequal conditions of access to resources. That is why, in order to identify and characterise production systems existing within an agrarian system, it is often essential to examine in great detail the conditions of access to productive resources for each category of producer: farmland, irrigation water, means of production and workforce, but also conditions of access to the market and information. Indeed, it often happens that conditions of access contribute significantly to influencing farmers' choices and therefore to explaining certain aspects of the productive combination being implemented. The case of farmers subjected to share-cropping, i.e. forced to give half of the harvest to the landowner in exchange for accessing the land and part of the means of production, gave rise to extensive literature and animated debates on its consequences as regards investment and technical progress.

In many ex-forest regions of intertropical America taken over by extensive cattle breeding frontiers (see Box 3.2), agronomical reasoning is affected by the very frequent association of crop cultivation such as maize with forage grass, a cropping association which is not only absurd since both species are in strong competition, but also incomprehensible as long as researchers ignore the social relations (a sort of share-cropping) which land-clearing farmers are subjected to, and the obligation they have to establish temporary grassland for the cattle breeders succeeding them the following year (Cochet 1993).

Methods for accessing land resources vary worldwide (right of way, right of collection, right of user granted free of charge, verbal renting, lease, shared access, private property) and cannot be limited to the mere public/private dichotomy.[15] The plurality of these methods, as well as the potential superimposition of different types of rights over the same area, does have an influence on production systems implemented by any category of producers, just as it sometimes impacts considerably on the distribution of value added and, therefore, on the income obtained *(infra)*. Understanding these methods is a compulsory step in the study of production systems. The same applies to conditions of access to collective resources (communal lands in particular) and the modes of regulation (collective actions and institutional arrangements) which are often associated with these.[16]

The same applies of course to irrigation water access and related regulation modes in many agrarian systems. In the Ecuadorian Andes, for example, agrarian history and contemporary political struggles are dominated by conflicts related to irrigation water: Indian communities who were dispossessed of this resource by major landowners, have since then been trying to recover their rights by any means (Gasselin 2000). How should one apprehend the operating logic of farms and their progressive trajectories, without having to fathom the relations of power and domination which are sometimes internal to the actual farming communities, and which restrict the field of possibilities to that extent?

Many examples could also illustrate how technical choices are, for a large part, conditioned by access to capital and therefore by the social relations in which farmers are involved. The difficulties encountered by farmers in accessing capital across the world, and the increased scarcity of credits granted to the smallest farmers in particular, gave rise to a new form of share-tenancy in many regions of the world:

[15]Sub-Saharan Africa offers many examples of multi-access to resources and potential superimposition of different types of rights on the same space. On this theme, and on the importance of taking into account these questions in development projects, see Le Roy et al. (1996) and Lavigne Delville et al. (2000) in particular. Concerning Latin America, see Aubron (2005, 2006) and Jobbé-duval (2005a, b) in particular. Several Internet sites, specialised in land ownership issues saw the light these last years. See the website of association Agter (Association pour l'improvement de la gouvernance de la terre, de l'eau and des ressources naturelles) on www.agter.asso.fr and the Landportal site in particular (http://landportal.info/).

[16]The outline presented by Elinor Ostrom in 1990 is of course a reference on this issue.

reverse tenancy, which concerns share-crop contracts as with classic share-tenancy, but in which it is not the holder of the land who is in a strong position, (as is the case for landless share-farmers who contribute their workforce to the production process) but, on the contrary, the share-farmer who, although without land, lays down the law because he brings in the capital required for the production process (i.e. all the fixed assets and sometimes the cropping costs). These new social relations were generally developed more recently in contexts where (1) land property was very largely distributed within the framework of agrarian reforms in particular, where (2) the beneficiaries of agrarian reforms did not have sufficient access to means of production, and where (3) liberalisation policies facilitated and made this type of contractual relation possible.[17] In this case, it is frequent to see these new "share-farmers" – holders of capitals and privileged access to the market – offering this type of contract to many small landowners totally devoid of means of production, and who sometimes become farm workers on their own land. Many examples of reverse tenancy can be given. Several have been described in Mexico (see Box 3.2).[18]

The many forms of "formalisations by contract" offered to farmers today, whether we are talking about a purchase agreement in return for adhering to specifications, or about taking over the production process entirely in exchange for a lump-sum payment, illustrates the abundance of methods for accessing resources in which farmers can find themselves involved for better or worse, with the combination of implemented production factors and the production system being often conditioned in a determining way by the paths taken.

The technical decisions taken by farmers are also influenced, and sometimes even imposed, by the reduced range of production means offered to them at some stage by training or "development" organisations, and therefore by economic and political choices largely beyond them. An example illustrating the weight of these "training techniques"[19] and political choices on the trajectories followed by production systems, is found with the development, in France in particular, of the maize-soya model for feeding dairy cows (Garambois 2011).

Production systems implemented by the farmers of a given region, cannot be apprehended as such without understand the complex fabric of social relations in which they are involved, especially in order to gather the production factors needed. In this sense, production systems are indeed subsystems of the agrarian system. Understanding them requires understanding the agrarian system just as the latter

[17]There are cases where this type of "reverse tenancy" is very old, as in Ethiopia where, for a long time, the holder of the harnessing equipment has been in a position to take lands as "share-farming" under very advantageous conditions to him, to the extent that in certain regions of the country, it is indeed access to the harnessing equipment (or fixed asset) which determines more than anything else real (not formal) access to land.

[18]See for example, in the case of Mexico, the works of Colin (2003) and those of Cochet et al. (2010a) on share livestock farming.

[19]Geographer Pierre Gourou was the first to define this term (Gourou 1973). In opposition to production techniques, "training techniques" designate all social relations of production, urban-rural relations and those which the public authorities can maintain with the farming community.

cannot be apprehended without a detailed analysis of these constituent subsystems (in this regard, see the issue concerning the combination of scales of analysis dealt with further on).

Production System Versus Activity System: A Badly Formulated Question

In many situations, and this is not a novelty, family strategies go beyond simple agricultural activities and can only be understood in the light of vaster strategies. It is then understood that the production process stricto sensu, studied with the help of the production system concept, is not meant to explain everything. What is more, it even often happens that the logics sustaining agricultural production systems cannot be apprehended without referring to "a *metasystem, called activity system, which constitutes the real domain of coherence of the farmers' practices and choices*" (Paul et al. 1994).[20]

However, the fact that farmers hold multiple jobs is not a new thing. Chayanov had already theorised about it in 1923, in his own way (Chayanov 1966). In the West-European context for example, and not so long ago, farmers were also carpenters, cartwrights, ropemakers, dairy men, cheesemakers, butchers, delicatessen dealers and traders... but were also, for all those whose farm did not keep the family workforce busy throughout the year (or the day), dressmakers, hawkers, chimney sweeps, woodcutters, etc. Multiple job holding was the rule rather than the exception (Mayaud 1999). It is rather the contemporary specialisation of farmers towards the production of non-processed basic foods, and the increasingly systematic call for inputs and tools not fabricated on the farm that has considerably reduced farmers' multiple job holding (*supra*). Similar findings could be made in different historical and geographical contexts. In many pre-colonial societies of Sub-Saharan Africa, for example, it is clear that multiple job holding was generalised among rural people: fishing, hunting, additional gathering of forest products, exchanges and long distance trading (e.g. kola nut). The fact that artisans such as blacksmiths were specialised, did not prevent them from also practising agriculture, which was made easier to them thanks to elders from founding lineages who gave them easy access to land ownership, to actually ensure that they would settle in the village. It is the colonial administration that was behind the brutal reduction in the mobility of populations and their multiple job holding, by confining individuals (i.e. administrative system applying to indigenous populations in pre-1945 French colonies, grouping of housing, forced labour of indigenous people) and by prohibiting access to certain areas/resources (Verdeaux 1997).

---

[20]In Anglo-Saxon works, this notion is tackled in terms of *rural livelihoods* in particular. See for example Ellis (2000).

Very few studies have dealt with residual multiple job holding at the end of specialisation processes, firstly because all those who practiced multiple job holding were not always considered as "real farmers", and therefore were not perceived as worthy objects of study and secondly because anything that was "external" to the production system stricto sensu, particularly production and trading relations which farmers could become integrated into, was rejected in the undifferentiated jumble of the "economic environment".

Today, and while we are witnessing in many regions of the world a redeployment of multiple job holding, it appears necessary to distinguish different versions of it, i.e. those relating de facto to the semi-proletarianisation of farmers or to generalised precariousness (the run-for-your-life scenario to ensure survival), from those leading to an increase in the standard of living and the realisation of productive investments (Dufumier 2006), or to the constitution or maintenance of an estate for retirement purposes. In the first case, it is because the farm income is not enough to survive on that multiple job holding is so developed; the idea being then to complement an income "to make ends meets". In the second case, multiple job holding can be "structural", the agricultural production system is then only an element of it. Farms, which can be very small (sometimes reduced to a garden-orchard) and produce only a limited income, are not for all that threatened with disappearance, because they are stakeholders in activity systems that include international migration.[21]

Anything that relates to the activity system and can help explaining the why and how of the productive process in agriculture (particularly its maintenance, even though the conditions for its intrinsic profitability are no longer met), must be examined very carefully. Taking into account these *other* activities in the study and understanding of agricultural production systems is essential. Appointing the family workforce to different activities (duration and season) depends on the farmwork planning and opportunity costs allocated to one work day or other on the farm, just as external income opportunities can bring the farmer to modify his timetable accordingly. This is where the concept of activity system takes on its full meaning, with the combination of activities generating income, social security, social links, estates, etc. On seeking to understand "what people do and why", there is no doubt that the limits of the production process sensu stricto are not enough to explain this, and that it is indeed at the level of the activity system that, *sometimes* (although not always), things make sense.

However, although the farming activities of rural households are not exclusive, or might not even be the main activities, they are still being organised coherently, all the more since it is often at the level of the communal territory, where the family unit is settled, that *operational* constraints are the tightest. The farm, even when it is reduced as an extreme case to a *residue* that only produces a limited share of the income (home consumption), like the patches of land formerly granted to the workers of Latin American *latifundia* or the co-operators of soviet *kolkhozes*, can be seen and analysed with the help of the production system concept. Of course,

---

[21]Many examples of these situations could be found in the Mediterranean region and other places. See for example the works of Adrian Civici on Albania as well as Biba and Pluvinage (2006).

the pertinence of this concept, even in these situations, should not dispense us from giving as much care to the *systemic* study of other activities developed within the family, activities that are not necessarily productive in the sense given to the *productions* of the production system, but that provide an income, create social links and are parties to social welfare mechanisms etc. The production system can then be perceived as a subsystem of the activity system.

Conversely, in the case of multiple job holding for survival reasons, it is more the fragility of the production system, its results which are too uncertain and its crisis, that explain the combination of activities established by the family to ensure its survival.

While we use the production system concept to describe and understand a group of farms, it is not always easy, however, to pinpoint one or several "other activities" that would be more specifically or systematically "associated" with the production system. This is the case when, for example, a farm type too small in size and with a family workforce not used to full capacity, is *characterised* by the sale of part of the workforce to the outside for working during peak periods (at neighbours' for example).[22] But it is not uncommon that labour opportunities are subjected to many contingencies, so much so that ensuring the meaningful coherence of a typology of agricultural production units (each type being represented by a production system) on the one hand, and a typology of activity combinations established by the family (activity systems) on the other, creates problems. Whence being tempted to go back to the activity system of each family, taken separately, and to go back to a sort of methodological individualism, meaning that each family would be considered as an activity system with its own production system (going back to the Osty conception of the production system *supra* ...).

The Cropping System and Livestock System Concepts

Here is another expression, another concept which is used to mean many different things. For a long time, geographers used this expression (and still do) to define the type of cropping plan of one "agrarian structure" or another characteristic of a given region. Moreover, some economists being less sensitive to the technical aspects of agricultural production and, in that sense, to the *operational* imperatives of the production system, sometimes compared the cropping system to all the cultivations carried out on the farm (which does not always form a system at that level of analysis), or even to all farm productions, with the production system in this case designating the simple combination of production factors leading to these productions.[23]

[22] A relationship which is still characteristic of the *latifundia/minifundia* relations in many regions of the world and which defines an "activity system".

[23] For example: Badouin (1987).

Faced with so much confusion, agronomists defined this concept in an accurate and truly systemic manner (unlike the different combinations designated under this very term by geographers and economists), and made of it an efficient tool for understanding what takes place "in the field".

Twenty years or so before Michel Sébillotte, Professor at the Institut National Agronomique of Paris-Grignon, gave the cropping system concept its current substance, Belgian agronomist Pierre de Schlippé had examined in great detail Zande manual agriculture (Congo), had classified "field types", and given this term the following definition (1956): "The field type is a structural concept: it covers the combination of a certain number of plants cultivated either as mixed cropping when sowing periods are simultaneous or successive, either in succession during the same season, or still in mixed sequence. Furthermore, it relies on a specific ecological context and, thirdly, it is characterised by a specific cultivation method which is a function of a specific farmwork planning" (de Schlippé 1956a: 100). He continued: "In order to have a perfect understanding of the Zande agricultural system, it was necessary to divide the subject in two, elements and structure, in the same way as a grammarian studies on the one hand morphology, the description of words, and on the other syntax, the description of sentences. We could compare the description of ecological formations, cultivated plants, tools and cropping methods to that of words, and the description of field types to that of sentences. Just as the meaning of a speech only becomes comprehensible when words are articulated into sentences, the why of an agricultural activity only becomes clear when its elements come together in the reality of the field" (idem), a definition that was surprisingly modern in the colonial context of the time. De Schlippé went even further: "Each field type has a name, which does indicate that it is considered as an entity, a concept, a model of behavioural norms, in short, a social *medium*" (de Schlippé 1956b). Far from the largely dominating technicist conceptions of the time, de Schlippé, by introducing the notion of cropping system, gave it an eminently social dimension.

With Sébillotte, the *cropping system* concept took on a more technical substance: it did not apply to a crop but to a plot (or a set of plots) cultivated in a certain way by the farmer. As such, it included the crop(s) planted (potentially as mixed cropping), crop sequences, and all the techniques applied to them following a specific organisation (i.e. a *crop management sequence*) and under given soil and climate conditions (1976). For a land surface treated homogeneously, the cropping system is defined through the crops planted with their sequence order and crop management sequences (logic and ordered combination of cropping techniques) being implemented.[24] Compared to the "field types" which P. de Schlippé was interested in, the concept defined by Sébillotte has the advantage of explicitly introducing the idea of temporal sequence, where it is possible for several "field types" to be inserted in the same crop rotation and, as such, to be part of the same *cropping system*.

---

[24] As to the crop management sequence concept, it refers to "the logic and ordered combination of cropping techniques making it possible to control the environment and to get a given production out of it" (Sébillotte 1974).

Observing and reading the landscape at first reveals *field types*, before any *cropping system* can be suggested; for in the end, the succession of different types on the same space is what can be analysed in terms of cropping system, and with a certain flexibility actually, because the successions in question are not always regular and repetitive, far from it. What is at play at the level of the cultivated plot, the cultivated plants and weeds developing in it, the conditions in which the whole thing takes place, the way it was dealt with, as well as the *history* of the plot, all this forms a *system*, or at least it is suitable to analyse it *in terms of system*, as one would a herd of domestic animals (*infra*).

On an equivalent scale of analysis, the *livestock system* is defined at the level of the herd, and integrates aspects relating to the herd structure (genetic characteristics, population pyramid, sex ratio etc.), its feeding and the corresponding forage calendar, as well as herd management (movements, reproduction and care among other things) (Dollé 1984; Lhoste 1984; Landais 1992). The livestock system can also depend on a number of practices: aggregation (constitution of units or batches, groups of animals that will be treated specifically according to their sex or age category, and that are related through animal flows), management (reproduction, health and feeding), farming (taking milk, wool, meat and more from the herd), and renewal of the herd (reform, selection of young animals or purchase for renewal purposes) (Landais and Balent 1995). Closely related in space and time, all these livestock farming practices must also be analysed in terms of system, with feeding very often representing a keystone.[25]

The analysis of a cultivated space in terms of a cropping system, or that of a herd of domestic animals in terms of a livestock system, includes of course many encompassing elements already encountered at the level of analysis of the production system immediately above, such as tools and workforce, which still represent a subsystem of the production system. Except for fairly rare production systems that include only one cropping or livestock system, it is indeed the combination of different cropping and livestock systems that, again, form a system at the level of the entire farm, at the level of the production system.

3.4 Combination of Scales of Observation, Analysis and Comprehension

At the regional level, geographers and agro-geographers were most likely the first ones to use the notion of agrarian system. As to the production system concept, it was and is used at the same time by economists, agricultural economists and agronomists, with the concept of cropping system elaborated by agronomists. But it is the combination of these scales of analyses and different concepts, the

[25]For a review of the use of this concept in Europe, see Gibon et al. (1999) in particular.

encompassing *agrarian system* in particular, that best reflects the originality of comparative agriculture as a comprehensive approach to world agricultures.

Space being the very object of geography, or at the very least a privileged dimension of analysis, the issue of choosing a scale of analysis and that of the potential combination of several scales were tackled by geographers. After a period where geography was dominated by regional approaches (between two wars and until the 1960s–1970s), there was a craze for larger scales (in the geographical sense), particularly the *terroir* or village territory,[26] before returning, more recently, to a smaller scale, that of the national and political context, i.e. the country. Out of all these spatial layers fitting into one another, only one was often adopted as privileged scale of analysis, while the others served as counterpoint. Jean Yves Marchal, on the contrary, proposed a "telescopic" change of scale, meaning that depending on the observation made on the ground, one should move up or down scales.[27]

Comparative agriculture favours the usage of a "telescopic" change of scale, particularly between the three levels of analysis privileged by us, i.e. that of the plot or herd for examining practices, that of the production unit or farm for integrating different cropping and livestock systems, and that of the (more or less vast) region or country for the pertinent application of the agrarian system concept. These are not simply three different spatial scales fitting into one another, but represent also and especially three levels of interdependent functional organisation. Of course, beyond these scales, the country, the subcontinental region or the world are all levels of analysis that cannot be escaped, because the competition affecting nearly all farmers across the world is so strong and immediate (*supra*).

Tackling these different levels of analysis and switching from one to the other as needed, does not prevent any form of hierarchical organisation, particularly because of the need to address *complex issues* starting from the *general* before studying the *particular*, as suggested by Marc Dufumier, i.e. "carrying out an analysis in successive stages, by starting with vast and encompassing levels of perception (the entire world, a country, a region . . .) to end with smaller and more specific levels (farms, plots, herds . . .)" (Dufumier 1996: 56).

Combining different scales of observation and analysis equates also to rejecting the idea that any problem can be apprehended, and a fortiori resolved, on only one scale of analysis. "Agronomic" logic (crop type, crop sequence, the "carry-over effect of the preceding crop" and the "sensitivity of the next crop", the nature and layout of cropping operations applied to each crop etc.) must be tackled in terms of systems at the level of the plot, but understanding it, explaining farmers' choices and practices also need to be pursued at the operational level of the combination of different cropping and livestock systems, i.e. at the level of the production system. Likewise, if all livestock farming practices developed around a herd of domestic

[26]With, for example, a series of studies on African *terroirs* conducted by Sautter and Pélissier (op. cit.) and many studies conducted on French villages. In this regard, see the outline presented by Bonnamour (1993).

[27]J. Y. Marchal, quoted by Bonnamour (1993: 117).

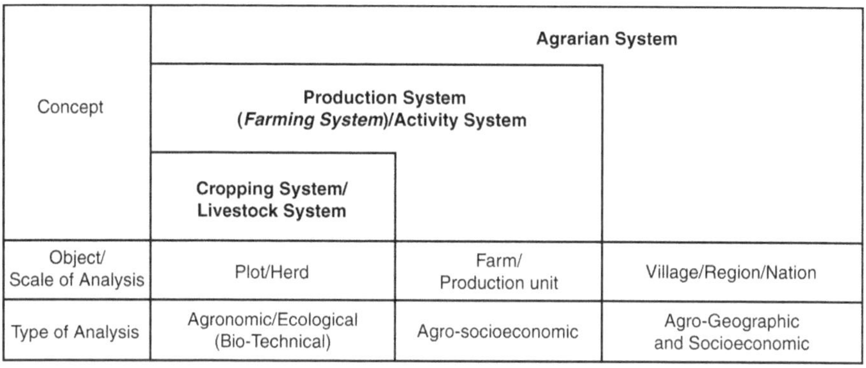

Fig. 3.1 Objects, concepts and conjoining scales

animals need to be analysed in terms of the livestock system, explaining farmers' choices and practices do not have to be pursued at this unique level of analysis, but can also be at the level of the production system (Cochet and Devienne 2006, op. cit.).

What is true about the to and fro between the *cropping system* and *production system* levels for example, is also true between the *production system* and *agrarian system* levels, between the agrarian system of a given region and those developed on the other side of the world, because the interweaving and interdependence of all forms of agricultures are so strong today (Fig. 3.1).

3.5 Between Life Sciences and Social Sciences, a Delicate Positioning of the "Agrarian System"

The agrarian system concept is complex and, let's face it, demanding. This complexity is a reflection of the reality it seeks to describe, i.e. a rural society or the agricultural sector of a society; what else could it be? Its complexity comes from the fact that combining very different scales of analysis on the one hand, and expressing the range of relations linking the technical sphere (exploited ecosystems and their functioning) to the social sphere (a productive social system) on the other, are demanding. Understanding the biological mechanisms governing the functioning of ecosystems and understanding the technical processes leading to their progressive artificialisation requires, as already seen, that knowledge and methods peculiar to life sciences in general, and agronomy and agricultural technology in particular (*infra*) be mobilised. But "then the agrarian system cannot be considered as a simple technical system of agricultural practices, nor can it be reduced to the sole distribution structures of lands intended for agriculture. The idea is, on the contrary, to analyse jointly the transformations of the agricultural techniques and the modifications intervening in the social relations, not only at the local level, but also at the national and international levels. This is what differentiates research conducted in terms of agrarian systems from works carried out as regards *farming systems research (FSR)* in Anglo-Saxon countries" (Dufumier 2007: 8) (*supra*).

Right from the 1970s and 1980s, the fact that the *system* approaches developed by Anglo-Saxon researchers, within the framework of the CGIAR in particular, did not really include the historical dimension of agrarian dynamics or farmers' technical choices as a reflection of their social and political context, explains why, on the one hand, the approach in terms of agrarian system evolved in parallel rather than jointly with systems grouped under FSR and, on the other hand, the agrarian system concept was not popular in the United Kingdom or in the United States of America.[28]

Indeed, it seems that in the Anglo-Saxon academic world, two families of approaches have been sharing all research dedicated to agriculture and the rural world. All those claiming to adhere to FSR focused on the study of technical processes in terms of systems, at the level of the agricultural production unit in particular, and which were often conducted by agronomists or their colleagues claiming to be followers of the sciences and techniques of agriculture. Farm typologies were then elaborated according to previously identified "differentiation criteria". Privileging the study of "systems" and their functioning at "t" moment (today) with a view to elaborating technical recommendations, it made little way for understanding long term processes, history, conditions for accessing resources, the distribution of value and its consequences, social relations as well as differentiation mechanisms and, finally, conditions for the integration of the farming community into society at large.

During the same period, while they were less concerned with explaining in detail the systemic nature of production processes, which was left to agronomists and economists adhering to FSR approaches, social approaches to agrarian issues led to a proliferation of studies as part of peasant or agrarian studies (Bernstein and Byres, op. cit.). Implemented by researchers in social sciences by referring in particular to political economy (or agrarian political economy), to sociology and history, these research works actually emphasised aspects which FSR did not address much or at all. Unlike previous approaches, these gave more than their due to social dynamics, history as well as the economic and political context into which farmers' practices and relations linking them to society as a whole fitted. They were putting forward social differentiation internal to rural societies, social relations and the role which integrating the rural societies of the South into the market played in the increase of inequality. On the other hand, peasant studies only rarely called upon the concept of "system", because the technical process as such was rarely central in the analysis. Moreover, there was a certain distrust vis-à-vis the approach in terms of systems, insofar as finding the characteristics of the system, its "balance",

[28]To the extent that the term "agrarian system" is not even mentioned is Anglo-Saxon dictionaries on human geography (see for example Johnston et al. 2000). Conway's approach, on the contrary, is indeed at the level of the agro-ecosystem (1984).

"internal coherence", "retroactions" and "regulations" inherent to the very notion of system as well as its "reproducibility" postulated as such, seemed incompatible, in the eyes of these researchers, with the revelation of internal conflicts, tensions and differentiations, social relations of production and exchange, periods of crisis and reconstitution, and therefore historical dynamics.

It is from this abundance of works relating to "peasant studies" that a new theme seems to have emerged in the USA from the 1980s onwards, a federative theme taking jointly into account physical and human factors in the degradation of the environment (Blaikie 1985), a theme around which *political ecology* (Peet and Watts 2004) was going to acquire an identity of its own. By postulating that, initially, ecosystem degradation processes were above all social and political, and by questioning the validity of the environmental policies conducted by the public authorities as well as the strategies and resistance caused by these very policies, political ecology led to the rediscovery of farming practices denounced in high places, their own logic and coherence. Following on these approaches, *environmental history* recently opened a way to examine further their historical dimension, thereby highlighting the complexity and dynamic of the relations between nature and society and, more specifically, between farmers and the ecosystems (i.e. between intensification and environment) in which they live Tiffen et al. (1994), Fairhead and Leach (1996), McCann (1995, 2005).

Comparative agriculture, on the contrary, tried to reconcile these two types of approaches around the agrarian system concept, and to favour their cross-fertilisation: the systemic approach of productive processes on the one hand, and an in-depth understanding of their insertion into society and long term agrarian dynamics, on the other.

Today, the comparative agriculture approach shows perhaps more similarities with American political ecology, because it postulates that agrarian dynamics result from the evolution of the relations between nature and societies, and from their expression at the interface between biotechnical process and social phenomenon. The Agrarian Studies programme, developed in the United States under the supervision of James C. Scott in the 1990s, was indeed built in reaction to the ahistorical nature of most studies and research works conducted on the "development" and "modernisation" of agricultures and rural areas. It was based on the two following postulates: (1) Relying more on collecting local knowledge and practices established by farmers, and therefore putting the local scale field work back at the centre of the analysis, and (2) implementing an approach which is comparatist and at the same time based on a corpus of multidisciplinary knowledge (Scott and Bhatt 2001).[29]

[29]"Every inquiry into rural life, whenever and wherever situated could be illuminated by work addressing similar problems in quite different contexts" (p. 2), with each research work to be "deeply touched by comparative reading across regions, across historical periods, and across disciplines. They are, in other words, analyses conducted in the shadow of a broad interdisciplinary encounter" (p. 3).

Through the importance given to the historicity of accumulation and differentiation processes, through the careful study of social relations, through the emphasis of field work and data collection, through the building of progressive theories on the multiplication of case studies while keeping away from global theories, through its holistic preference and systemic approach, and finally through its openness to other social sciences called upon as needed (*infra*), comparative agriculture showed a few similarities with certain so-called "heterodox" approaches, "classic" institutionalist approaches in particular. In this regard, Jean Philippe Colin and Bruno Losch pointed out certain proximities between what they called "French Africanist rural economy", to which comparative agriculture contributed, and classic American institutionalism (1994).[30]

The agrarian system concept is perhaps too complex and encompassing, or its application too difficult to implement within the narrow framework of a mono-disciplinary research project, or that of an extension operation, for its usage to be widely spread within a scientific community which is excessively specialised and little inclined to seeing "hard and social sciences" coming together. Could it be that its explanatory ambition is disproportionate and its usage limited to that of a kaleidoscope?

Many gave up. More than 30 years after the creation of a department specialised in the study of agrarian systems at the French Agricultural Research National Institute (INRA) (in 1979) and the International Centre for Research and Development (CIRAD[31]), the reflux of the approach indicated that the agrarian system concept, and with it the regional scale on which it was most pertinent, would be somewhat neglected today.[32]

The fact that some abandoned this concept and this scale of analysis of agricultural dynamics left a void, which meant that everything beyond the farm was still too often rejected in the undifferentiated magma of "economic environment", while relations between these "external" elements were complex and therefore had to be analysed more than ever in terms of systems. Moreover, do not the strong return of the "local", "landscape" and "territory" concepts in environmental approaches, and the increasingly felt need to generally understand problems and the indissociable nature of the "technical" and "social" aspects, require to focus once more on this scale of analysis and to apprehend it *all* so as to understand parts of it?

[30] See also Colin (1990).

[31] French agricultural research organization working for development in the South.

[32] To the extent that the INRA's Agrarian System and Development Department (SAD) changed its name, if not its acronym.

Chapter 4
Diachronic Approach to Agrarian Systems

Thanks to the specific concepts it was able to develop (the agrarian system concept in particular), comparative agriculture is in a position to shed an innovative light on the old and contemporary transformations of agriculture, which have sometimes been interpreted in terms of *crisis* or *agricultural revolution* as will be seen, and as such is able to take part, jointly with other disciplines such as History of course, but also Archaeology, Ethnobotany and Historical Technology, in the creation of knowledge and understanding of the key phases of the evolution of world agricultures.

The compared study of agrarian systems, in a synchronic approach, must go together with their diachronic study, i.e. their succession and sequence in time. "The evolution, breakdown, break-up and reconstitution dynamics of agrarian systems results from the contradictory movement of the different levels of the system, with their respective evolutions falling within a suitable and more or less autonomous dynamics" (Kroll, op. cit.: 12). In this movement, the notion of "agricultural crisis" and "agricultural revolution crisis" appears fundamental. Insofar as we can consider that crises and agricultural revolutions often interrupt the passage from one agrarian system to another, we can say that these notions are indissociable from that of agrarian system.

History gives us a number of examples where periods of relative stability and/or continuous growth, or on the contrary periods of endless misery replace periods of unrest, regression then deep transformations and rebirth. Pinpointing these major discontinuities punctuating the history of agrarian societies, analysing them and making them intelligible are all essential tasks of comparative agriculture. Understanding an agricultural or agrarian system crisis, and thinking about the means and requirements to be met with a view to overcoming such a crisis or anticipating its advent, are all major issues of this discipline.

H. Cochet, *Comparative Agriculture*, DOI 10.1007/978-94-017-9828-0_4

4.1 What is an "Agricultural Revolution"?

The Concept of "Agricultural Revolution"

The term was first used to designate the deep transformations of English agriculture in the eighteenth century, particularly in relation to the enclosure movement. Its usage was then extended to include the significant transformations of West-European agriculture in the nineteenth century, notably in France, which was characterised in particular by the cultivation of fallow lands with fodder roots and tubers and forage crops. This agricultural revolution was described and analysed by historians such as Marc Bloch (1931) and Michel Augé-Laribé (1955). In *Les caractères originaux de l'Histoire rurale française* (The Original Features of French Rural History) (1931), Marc Bloch wrote: "It became a habit to call 'agricultural revolution' any major disruption in agrarian techniques and customs that, throughout Europe, at different dates according to country, marked the advent of contemporary farming practices".[1]

The use of the word "revolution" was sometimes contested, in that the transformations in question were often devoid of the suddenness and rapidity suggested by this term. Marc Bloch declared in this regard: "The term is practical, [...] it emphasises the extent and intensity of the phenomenon [...]. A revolution perhaps, [...] although hardly an unprecedented jolt after centuries of immobility. Could it be a sudden mutation? Not the case either. It spread over many years or even centuries".[2]

In fact, the slowness and hesitations of the process take nothing away from its revolutionary nature. On the other hand, one can wonder about what differentiates a simple set of agricultural and agrarian transformations from what is rather appropriate to call "agricultural revolution". In *L'Histoire de la France Rurale* (The History of Rural France), G. Bertrand wrote: "The passage from one 'model' of agrosystem to the other corresponds to a mutation in the relations between rural society and its ecological environment. Analysing these mutations is essential. It presents, from an ecological point of view, the much debated issue of 'agricultural revolutions'" (Bertrand 1975). For M. Mazoyer "since any agrarian system, any mode of exploitation of the environment has a limited production capacity, as long as this capacity has not been reached, agricultural development can consist in extending this system, in fully using its specific means and in refining its utilisation in the entire exploitable space. This limit being reached, agricultural development can only continue by changing the agrarian system, i.e. an agricultural revolution involving a change in the quality or nature of the production process (other tool system, other source of energy, other cultivated ecosystem, other mode of artificialisation of the environment, etc.). In both cases, i.e. (1) the extension of a

[1] p. 201 in the Armand Colin edition of 1976.

[2] idem.

pre-existing agrarian system and (2) the establishment of a new agrarian system, although especially in the second case, improvement of the production process supposes prior changes in the working and trading relations, in the institutions and ideas governing this process and potentially hindering its progression. As such, an agricultural revolution does not consist in merely extending the pre-existing farming method, on the contrary, it includes a deep qualitative change in the production process that affects the nature of the cultivated ecosystem, the force or quantity of energy used, the power of the tools used, labour productivity etc."(Mazoyer 1987).

It is essential to recall that an agricultural revolution is not reduced to a technological revolution or to the simple adoption of new or foreign techniques, tools and knowledge. In the case of the agricultural revolution of the eighteenth and nineteenth centuries in Western Europe, forage crops were not a novelty, nor were fodder roots and tubers that were already known and cultivated in gardens for a long time already. The whole tool system was already known and available since the Middle Ages (harnessing, plough and harrow, scythe, carts and waggons). What was new was the combination of social, economic, legal and political conditions, thanks to the Revolution in particular, which made change *possible and necessary*. "It was a complex agricultural development, inseparable from the development of other sectors of activity, and whose conditions and consequences were ecological, economic, social, political, cultural and juridical, much more than technical" (Mazoyer and Roudart 2006: 332).

The use of the agricultural revolution concept was first limited to the European agricultural changes that took place in the eighteenth and nineteenth centuries; it was then extended by M. Mazoyer to include other geographical and historical contexts. By describing the agricultural revolution of the eighteenth and nineteenth centuries in Europe as the "first agricultural revolution of modern times", he introduced the idea that others preceded it, with the very first being of course the Neolithic agricultural revolution that, although carried out in different places ("centres") and not in a synchronised manner, still remains a disruption of a comparable nature, at the technological as well as social, political and cultural levels.

The Four European Agricultural Revolutions

Concerning European agricultural history, M. Mazoyer showed very well how and why all four sets of transformations, described and interpreted by him, actually constitute agricultural revolutions through their scale as well as their technical, economic, social and political determinants. He showed that the development of agrarian systems based on fallowing and animal-drawn cultivation with the ard in temperate regions and their *ager/saltus* differentiation, constitute the "agricultural revolution of antiquity" and was an appropriate response to the problem posed by deforestation during that period (see Box 3.1). The development of agrarian systems based on fallowing and animal-drawn cultivation with the plow in the cold temperate regions of North-West Europe, during the Middle Ages, is then described as the "agricultural revolution of the Middle Ages". This revolution was characterised by the significant development of winter forage supplies, thanks to the extension of meadows where grazing animals were prohibited, transport means (carts) and the storage of forage (hay-barns); by the maintenance of larger herds and an increase in manure production; by the concomitant increase in cereal yields and the generalisation of the three-field system in Oceanic and Northern Europe; and finally by the use of more efficient ploughing tools (ploughs). The third agricultural revolution identified, already mention above, is that of the eighteenth and nineteenth centuries in Europe, which was described as the "first agricultural revolution of modern times". Finally, the transformations experienced since the end of the Second World War by agriculture in Western Europe and the United States, constitute the "second agricultural revolution of modern times" or "contemporary agricultural revolution". Its characteristics are well known: motomechanisation, specialisation, "chemicalisation", selection, (Box 3.1).

Agricultural Revolutions in Sub-Saharan Africa

In another geographical context, the history of agriculture in Burundi, in the central African Highlands, could be "summarised" with the succession of three distinct agrarian systems, where their passages from one system to another can all be characterised as agricultural revolutions (see Box 4.1). We can indeed show that the agrarian mutations which, during the seventeenth and eighteenth centuries, came with the introduction and development of maize and American beans, i.e. (1) the generalisation of farmwork planning with two cropping seasons per year, (2) the multiplicity of harvests and global increase in labour productivity, (3) the accumulation of cattle and meticulous recycling of animal manure,[3] (4) the development of social practices built up around the distribution of livestock, and (5) the acquisition of relative food security and the reinforcement of the State, all this does constitute an agricultural revolution, the first in that history (Cochet 2001, see Box 4.1). The "banana revolution" constitutes the second revolution and took place during the second half of the twentieth century. It was characterised by the development of banana plantations near dwellings which quickly became agroforest gardens, by the increasing complexity of crop associations on slopes, the multiplication of cropping cycles in former drained lowlands, as well as the development of coffeegrowing, all these transformations leading to labour-intensive multiple cropping enabling the Burundian rural communities to feed three times more people, while taking their place on the world market of quality *arabica* coffee (idem).

[3]The basis of an agriculture-livestock farming association that was much older than was generally believed for this part of the world.

In the case of Burundi, as mentioned above, let us note that the two agricultural revolutions in that part of the world were not characterised by a change in energy source – we are still in the domain of manual agriculture – or a change in implements (indeed, the iron hoe went through 2,000 years of agrarian history without experiencing any decisive modification). Rather, they were characterised by the available biological material and its usage: first there were beans and maize, new plants for which farmers transformed their old farmwork planning to ensure good harvesting, an astonishing example of rural farming innovation with the perfect integration of new plants into farmers' practices; then there was the banana tree, a plant that had been known for a long time in the area, but that only became generalised in the twentieth century, accompanied by a consumption pattern (banana "beer") which was original and at the same time particularly economical as far as fertility was concerned; and finally, sweet potatoes and manioc, as well as trees (avocado pear trees, mango trees, eucalyptus and lacewood among others), with increasingly sophisticated associations and combinations leading to the establishment of terrace orchard gardens, and to the emergence of entirely new agroforest landscapes. Carried out (almost) without calling on industrial means of production or fossil energy, these transformations were like an agro-ecological model. Apart from the biological material, the farm capital was also metamorphosed. The farmer's fixed assets, which included not only the limited implements but also the live components, were especially reinforced: first the cattle, then the banana plantations and accumulated fertility, all this after the accumulation patterns and social relations in use underwent a major change (idem).

Box 4.1: Crises and Agricultural Revolutions: The Example of Burundi, in the African Great Lakes Region (Cochet 2001)

Together with neighbouring Rwanda, Burundian rural areas constitute one of the densest centres of rural human concentration in the world. Average population densities exceed the 300 inhabitants per km^2, with some regions exceeding 700. Manual and almost exclusively rainfed agriculture relies on the organisation of species and cultivated varieties originating from all continents and combined between agroforestry, complex crop associations with two cropping seasons per year, coffeegrowing and the development of the lowlands. This labour-intensive agriculture, although making very little use of industrial means of production (implements, artificial fertilisers, pesticides and fossil energy), and which in this sense is agro-ecological, is the result of a long agrarian history: three agrarian systems succeeded one another in Burundi, with the passage from one to the other taking place during two major agricultural revolutions.

(continued)

Box 4.1 (continued)

The Old Association of Agriculture and Livestock Farming

During more than 1,000 years, sorghum and African millet (*eleusine coracana*), which are sowed at the beginning of the rainy season, constituted the basic food of Burundians. These crops were fertilised thanks to the cattle manure collected each morning in the enclosure where the cattle spent the night. Beyond the cultivated lands that were grouped close to the enclosure, vast grasslands were grazed by herds during the day. Recycling the night manure and spreading it onto cultivated lands represented the two stages of a lateral transfer of fertility for the benefit of the crops, and ensured the continuity of the system.

The Agricultural Revolution of the 17th and 18th Centuries

From the seventeenth century and during the entire eighteenth century, a real agricultural revolution came to disrupt this old agrarian system. With the massive adoption of plants of American origin (maize and *phaseolus*-type beans), and the fact that they were growing at the beginning of the cultivation season, the maize + beans (during the first cropping season)/sorghum (during the second cropping season of the same year) succession became generalised throughout the country. After a significant increase in the work supplied, farmers could from then on practice two cropping cycles per year. Net productivity per labourer increased considerably. Spreading the harvests, reducing the risks and substantially improving the food enabled the population to escape food shortages and starvation.

While this evolution made heavier demands on the mineral reserves of the soil, more efficient fertility reproduction mechanisms were established. Increasing cattle numbers was not sufficient on its own to meet these needs; it was also necessary to focus cattle manure on the new crop rotation with two harvests per year, which was more intensive and practiced close to the enclosure, while the plots further away, where only African millet was grown, were deprived of it. This is how the accumulation of cattle, and therefore manure, came to be at the centre of farming strategies, social practices and clientelist relations built up around the distribution of the livestock.

These transformations go beyond the mere domain of production and affect society as a whole, its organisational forms, the social relations of production and the distribution of value, its cultural, ideological and political blossoming with the consolidation of a centralised kingdom. That is why it is fitting here to speak of a real *agricultural revolution*, even if it took a while – several generations perhaps – for these changes to become generalised throughout the Burundian Highlands. The expansion of the population, stemming from this agricultural revolution, is behind the populating rate that was reached at the end of the nineteenth century, and which was exceptionally high compared to the rest of the continent.

(continued)

Box 4.1 (continued)

A Malthusian Crisis

Towards the end of the nineteenth century, the development process was abruptly interrupted and the country experienced a serious crisis (epizootic diseases, food shortages and starvation), which lasted until the middle of the 1940s. Prior to this series of disruptions, the cattle plague had played a decisive role. Indeed, cattle being at the centre of the accumulation system of the farms of the time, epizootic diseases resulted in farms suddenly losing their capital and in the transfer of fertility being interrupted, to the benefit of arable lands.

But the cause of this crisis was first of all linked to the generalised weakening of the cattle, provoked by overgrazing. With 70–75 inhabitants per km^2 at the end of the nineteenth century, the agrarian system had reached and even exceeded its own limits. Pursuing population expansion led from then on to the reduction of pastures, with parts of them being dedicated to ploughing. While everybody was trying to maintain or increase their cattle numbers to safeguard crop fertilisation, particularly on common grazinglands – that's the "tragedy of the commons" – the depreciation of the pastures/fertilised lands ratio led to generalised overgrazing, the weakening of the cattle and its collapse during epizootic diseases. Finally, the quantity of animal fertiliser available per cultivated hectare dropped sharply, leading to a drop in labour productivity and in the quantity of food available.

Although from every angle it was definitely more efficient than the previous system, the agrarian system with two cropping seasons stemming from the previous agricultural revolution also had spatial requirements, specifically as a result of the necessary association of cultivation with cattle farming. Because of its efficiency, it ended up reaching the Highlands where it blossomed. Although foreign colonisation and forced integration into trading were in no way related to the deep causes of the crisis and its triggering, they extended the crisis well beyond what would have been possible.

Banana Plantation and Intercropping: Towards a Labour-Intensive Multiple Cropping

Between the middle of the 1940s and the beginning of the 1990s, and although the population had increased threefold and by then neared 200 inhabitants per km^2, the country acquired and maintained self-sufficiency as far as food was concerned. While cultivated lands increased (to the detriment of pastures) by around 50 % during the period from 1950 to 1990, food production experienced a growth in the region of 150 %, i.e. three times faster. This remarkable dynamics was the result of the progressive and continuous intensification of production systems.

The multiplication of banana plantations and their extension was one of the clearest demonstrations of it. Planted also with varied fruit trees,

(continued)

Box 4.1 (continued)
multiple cropping represents the most successful cropping system in terms of value creation per unit area or per day of work and, together with coffee, represents the first source of farmers' cash income. In terms of fertility, banana plantations play a rather exceptional role, thanks to the specificity of the product sought after: banana juice intended for consumption or sale after being fermented into "beer". Where all crop and harvest residues (including banana skins and pressing residues) are returned to the soil, losses in mineral elements due to harvesting are practically reduced to naught while organic matter accumulates on site. The result is that, once established, this cropping system could be self-sufficient by not requiring animal or chemical fertilisers. The expansion of banana plantations explains in this way the decline in cattle farming, which it replaced in the end.

On the hillsides, beyond banana plantations, crop associations became almost systematic and progressively complex, accompanying the multiplication of cropping cycles on the slopes as much as in the lowlands.

The increasing complexity of associations, the development of banana plantations, the transformation of the fertility reproduction method, labour intensification and, finally, the doubling or tripling of food production: this is another agricultural revolution, occurring without any industrial means of production, fertiliser, pesticides or fossil energy, somehow giving us a general idea of an agro-ecological revolution.

Unlike the agrarian transformations of the seventeenth and eighteenth centuries that mainly led to an increase in labour productivity, the banana revolution led to an increase in production per unit area. By progressively replacing cattle, banana plantations eventually did without them and their supporting pastures. Moreover, the increasing complexity of crop associations on the hillsides, and the multiplication of cropping cycles in former marshes, led to a tighter management of the resource and to saving significantly on space.

The example of Burundi is a good illustration of how reconstructing the different agrarian systems succeeding one another in this region, and the use of the "agricultural revolution" concept to characterise and interpret the disruptions that led to their succession, address the issue of the population/resource balance by going beyond the simplistic debate between followers of the neo-Malthusian theses and those who refer to the model put forward by Boserup in her days (*supra* and Box 4.1).

The same applied in South Mali, in what was called the "cotton areas". Population growth and crop expansion did not follow the expected path, whereby the time period during which lands remained fallow was reduced and slash-and-burn agriculture accelerated, with all the negative consequences often denounced in such situations. The ongoing transformations, of a completely different nature, can there again be interpreted in terms of agricultural revolution and, consequently, better understood. Beyond fields located near houses, these gardens that were continuously cultivated for a long time, new spaces formerly integrated into the slash-and-burn agriculture/fallow land succession (bush then forest "fallow"), were from then on also cultivated every year, while the less favourable spaces of the village territory (hardpan plateaux and slopes in particular), also subjected to periodic slash-and-burn agriculture in the past, were allotted to grazing. The differentiation of a real *ager* and a forage providing *saltus* came with the development of livestock farming and fertility transfers for the benefit of the *ager*, thanks to the generalisation of the compost technique actually permitted by the development of animal-drawn transport. The intervention of the public authorities was crucial in this case, particularly within the framework of the extension and subsidising of new equipment (animal traction) for cotton growing, although the ongoing transformations of this agrarian system went well beyond the impact of the cotton growing operations strictly speaking (Bainville and Dufumier 2007).

Green Revolutions

By "green revolution", we usually designate the progress realised from the 1960s and 1970s in the agricultural sector of certain countries of the South (North-East of Mexico and Indian Punjab among others), a process based on the utilisation of varieties (rice, maize and wheat) with potentially higher yields, artificial fertilisers, pesticides and irrigation, in a context of high State intervention in agricultural matters. Strongly encouraged – and financed – by private foundations and Northern governments, from the Western Bloc in particular that saw in them a bastion against red revolutions in the context of the Cold War, green revolutions were imposed especially from the top, like so many advisory packages proposed to farmers. When all the conditions required for the externalisation of the yield potential of the new varieties were met, the result was often a significant increase in yields (Dufumier 2004; Cheyroux 2005; Devienne 2006). The increase in labour productivity was often more modest: the partial motorisation of tasks was sometimes developed (two-wheeled tractors and threshing machines), but the productive process remained partly based on human and animal energy. When new means of production stemming from the "green revolution" could be acquired by a very large section of the farming community, because shared access to land and irrigation water had been acquired previously, particularly within the framework of the agrarian reform, real development processes could then emerge, resembling agricultural revolutions.

But where the same advisory packages were transferred, often authoritatively and in a context of great inequalities as far as access to productive resources was concerned, thereby generating a significant increase in inequalities and in the exclusion of large sections of the rural population, can the productivity gains obtained, which incidentally are far more modest, and the changes occurring within these rural societies, be described as "agricultural revolution"?

Moreover, this "green revolution" remained inaccessible to the farming community whose cultivation could not be artificialised or homogenised in conformity with the requirements of the new varieties, in regions with rain-fed agriculture in particular (Dufumier 2004). On the other hand it contributed decisively, as did the contemporary agricultural revolution in Europe and the United States, to the generalisation of simple and standard advisory packages, based on a very small number of species conducted as sole cropping, and calling on increasing quantities of artificial fertilisers and pesticides, to the detriment of generally more complex systems calling largely on the "ecosystemic" functions that were prevailing until then.

Every Agricultural Disruption Is Not Automatically an "Agricultural Revolution"

A contrario, we could also give examples of transformations that do not constitute in any way an "agricultural revolution", notwithstanding their scale and consequences. This is the case, for example, of the transformations that accompanied the "discovery" of the New World: introduction of new vegetal and animal material (wheat, barley, broad bean, sheep, cattle, goats, horses and pigs), an entirely new tool system (iron hoe and mattock, swing plough and yoke, wheel etc.), an entirely new relationship with land ownership (private property), and unprecedented trade relations among others. Although everything could have contributed to the advent of a spectacular agricultural revolution, nothing came of it. It took two to three centuries for the farming communities of the Andes, Central America and Mexico to make this new heritage theirs and to derive a minimum of well-being from it, after having experienced the worst agrarian, political and demographic crisis probably known to humanity ... Who would speak of agricultural revolution to describe such a process? Who would speak of it to describe the chaotic development of the plantation economy in Africa, in the form of large domains during the colonial era, following spectacular and destructive frontier periods, or to describe current processes for the acquisition of agricultural assets by public or private investors in the countries of the South and the former Soviet Union?

Could true agricultural revolutions be first and foremost internal and endogenous processes, integrating exogenous inputs although only insofar as their assimilation comes to reinforce something or fulfil a shortage, crystallising new forces under favourable conditions met at last? In short, a fantastic *self-centred* agricultural development process, never imposed from outside, and during which a non-negligible majority of the population would see its standard of living increase? And a contrario, do not the sudden contact of two agrarian civilisations, the fact that one is dependent on or rejected by the other, the more or less authoritarian imposition of new means of production, of new ways of doing and thinking, constitute the very negation of the agricultural revolution concept? If the transformations of the agrarian system, deep though they are, do not lead to an increase in food production, an increase in the density and/or well-being of the maximum population allowed, an at least partial sharing of growth, in short a decisive and lasting progress, can we speak of agricultural revolution?[4]

4.2 Agricultural Crisis, Agrarian System Crisis

An agricultural crisis, or that of an agrarian system, is first of all the accumulation of different types of tensions, blocks and brakes, and/or the appearance of major contradictions ending up challenging the current accumulation process, its control methods, the social relations underlying its development, and sometimes even society as a whole in its very existence. But crises can vary in nature, resulting from different processes and must therefore be analysed accordingly. In this regard, we can mention for example:

Crises Reflected in the Irreversible Degradation of the Environment

In a first category, we could group together agrarian crises which, in ancient historic periods, led to the complete and final destruction of the societies in question: the hydraulic agrarian system of the Marib Region (Yemeni piedmont at the time of the Empire of Saba), the agrarian system of the classic Maya civilisation of the Lowlands of Yucatán (Mexico) and Petén (Guatemala), which combined cropping systems on slash-and-burn agriculture with ingenious irrigated gardens, the agrarian system of the Kingdom of Aksum in current Northern Ethiopia, based on cereal growing with ard (*maresha*), or that of Easter Island of which the collapse was recently popularised by Diamond (2005); all agrarian systems with a crisis manifested first through the irreversible degradation of the environment (deforestation of Yucatán and Petén around the tenth century, climatic depreciation of the Tigray Plateaux around the seventh and eighth centuries AD, floods and destruction of hydraulic infrastructures in Yemen in the seventh century), even if the underlying causes of these cataclysms, which are still not widely known, indicate that environmental factors are of course not the only factors involved ... Likewise, the crisis of forest agrarian systems, based on slash-and-burn agriculture, is often inescapable when population density exceeds the level permitted by the rate at which biomass regenerates during periods separating two phases of temporary cultivation.

[4]From that point of view, is the English agricultural revolution – as the first one to have been designated as such – really an agricultural revolution, if we measure the resource concentration phenomenon and scale of exclusion phenomena cause by it?

Closer to our time, the Sea of Aral drying out due to the overexploitation of the rivers flowing into it, is a strong reminder that this type of agrarian crisis can be current affairs, as are the potential consequences of climate change on the most vulnerable rural areas.

Endogenous Crises

Another type of agrarian crisis can be identified when spatial expansion, and likewise that of an agrarian system, encounters physical, geographical or territorial limits, revealing new tensions relating to the appropriation of resources and their exploitation, and provoking conflicts to the extent of questioning the social structure as a whole. For example, the new agrarian system of North-West Europe, stemming from the agricultural revolution of the Middle Ages, and based on cereal cropping with fallow lands and on the agriculture-livestock farming association (pastures of the *saltus*, meadows, etc.), reached several times its limits, with the increasing population being on several occasions destroyed by starvation and epidemics, every time the population limit and maximum population capacity of the agrarian system was reached (Mazoyer and Roudart 2006). The system is hindered by its insurmountable limits, with no underlying technical, social or political transformation taking place; in short, there is no new and more effective agrarian system. The same thing was illustrated in the Burundian example above (Box 4.1). The agrarian system with two cropping seasons, stemming from the agricultural revolution of the seventeenth and eighteenth centuries, was so efficient that, in the end, it filled up in men and cattle to the extent that its physical limits were finally reached and exceeded. It was indeed the crisis of this agrarian system, revealed and triggered by the cattle plague at the end of the nineteenth century, which opened the 50 years of agrarian crisis that had to be endured up until the middle of the 1940s. And although the elements of a brand new agrarian system were already being established since the beginning of the century, it was only during the 1950s, under new political conditions, that the crisis was truly ended, from the top. In both examples, we are talking about "endogenous" crises, i.e. the crisis of a mode of exploitation of the environment reaching its limits, the crisis of an old order that became impossible to maintain, and the crisis of the legitimacy underlying the social relations and the institutions binding them.

External Crises

A third type of agrarian crises is manifestly triggered by "external factors". The destruction of pre-Hispanic agrarian systems in Mexico and the Andean countries is a typical example of it. Another example is given with the destruction of agrarian systems based on mixed crop-livestock farming with animal traction that had developed in certain regions of South Africa, a destruction which was planned by the public authorities within the framework of apartheid. Lastly, the competition between farming communities highly unequally endowed in production means, and the drop in the relative prices of their products plunged many farming communities of the Third World into an increasingly deeper crisis... Sometimes still, it is the rapid and massive occupation of an important portion of the territory by non-local producers that provokes or amplifies the agrarian system crisis. The sudden decrease in the space for the reproduction of forest biomass and therefore the length of time that space has been lying fallow with trees, the (colonial or contemporary) perennial plantations in forested Africa, or the development of extensive cattle farming on the frontiers of intertropical America, are all examples of crisis-generating processes. There is no doubt that, today, the movement for taking control of vast agricultural lands by foreign investors in many countries of the South is also likely, insofar as the coveted lands are most often occupied, to generate very serious problems in the years and decades to come.

Political Crises?

In other situations, some elements come to disrupt or slow down ongoing transformations, prevent the development of the agrarian system, hinder its extension and prohibit it from using its biological and human technical means at full capacity. For example, what hinders the pursuit of agricultural progress in Burundi today, does not relate in any way to the fact that the population limit imposed by the productive capacities of the agrarian system have been exceeded, or even to the competitiveness imposed by more productive regions or countries. The causes of the crisis do not have to be sought either in actual farmers' practices, despite their repeated denunciation by development services and public authorities. On the other hand, today the ongoing intensification process (Box 4.1) which can still be pushed further, is slowed down and hindered in many ways: massive diversion of biomass for the exclusive benefit of coffeegrowing via compulsory mulching,[5] shortage of production means, relative pricing system making fertilisers and pesticides inaccessible, repeated and massive offences against the security of the tenure, all obstacles which make it increasingly difficult to pursue and deepen ongoing intensification dynamics since the 1950s. Today, the progress of the Burundian agrarian system is limited by the conditions in which producers are integrated into trading, and by the State/farming community relations in which they are involved.

A sudden change in economic policy can also be the cause of agrarian crises. The structural adjustment policies imposed by sponsors to developing countries in the 1980s, could illustrate this remark: dismantling of market regulation tools,

[5]Mulching consists in spreading, every year, on the coffeegrowing plots, a thick layer of fresh mulch which the farmer must find wherever he can, generally on the other plots of his production unit (crop residues, banana leaves and trunks, cut grass etc.).

withdrawal of farm input grants and equipment financing, shift in the balance of relative prices. Many agrarian system were weakened by these political changes, to the extent that they were sometimes plunged into a deep crisis, it being understood that, in this case, other factors were also frequently at play to stimulate the crisis.

Many-Sided Crises?

Finally, certain multi-sided crises can result from several convergent dynamics: the example of the political crisis affecting Ivory Coast at the beginning of the 2000s, with the agrarian matter being absolutely central to the crisis, is meaningful in this regard. As the leading world cocoa producer and, on its own, supplier of half the world production, Ivory Coast owes this spectacular development and its economic expansion during the 1960s, 1970s and 1980s, to the cultivation of the cocoa tree which, in the space of a few decades, swept the whole southern part of the country formerly covered with forests. This plantation dynamics was made possible by the conjunction of three series of favourable conditions: first of all the existence of a vast forest, still largely protected until the 1950s, which constituted an excellent "cropping history" for establishing cocoa plantations; secondly, a considerable input of labour, thanks to the launch of a vast movement for the migration of farmers from the Northern regions of Côte d'Ivoire then neighbouring countries, leading to the rapid expansion of the plantations; finally, the establishment of a particularly attractive agricultural and migration policy ("whoever plants owns the land", investment subsidies, guaranteed and relatively stable prices). The dynamics of this frontier can be explained with the cocoa production cycle, where a phase of rapid plantation production growth is always followed by a phase of aging and crisis, making it necessary to establish new plantations to the detriment of new forest areas. But as soon as the "forest differential rent"[6] ran out (massive deforestation), as soon as attacks from parasites multiplied (at the densification and aging rate of plantations), as soon as the remuneration of planters dropped under the double influence of cost increases and world market decreases, there went the economic, social and politico-ethnic consensus that had been the basis of this spectacular development.

The crisis which the agricultural sector of a country like Ukraine went through during the 1990s, soon after the fall of the Soviet Union, was also many-sided: the political collapse of a system, but also the sudden interruption of the networks for the supply and sale of products provoked by the country's brand new independence, the social and economic contradictions and blocks accumulated within the very kolkhozes and sovkhozes inherited from the former regime, the sudden removal of their capital and the rapid abandonment of animal productions, the collapse of employment in rural areas, etc.

---

[6]In this regard, see Ruf (1995).

4.3 Comparative Agriculture, Agricultural "Crises" and "Revolutions"

From this typological outline, we can see that agrarian crises can be very different in nature. Their identification and study in detail are essential to understanding actual agrarian systems. Research and identification of agrarian crises and transformations, potentially established as *agricultural revolution*, are an integral part of the research on actual agrarian systems. While the use of the actual concept of agrarian system requires to pause somewhere in order to describe its structure and functioning at a given period of its history, even if *in the end* its internal dynamics is what is being researched, the analysis of the transition periods during which the transformations giving rise to another agrarian system are established, is dialectically linked to this dynamics. While transformations are difficult to demonstrate without having hypothetically at least an *initial* and a *final state*, conversely it is the comprehension of the change that makes it possible to better understand the different successive states of the society under study. Moreover, agrarian systems cannot be considered as *stable* structures, and all agrarian transformations as possible courses to shift from one system to another; internal dynamics, whether or not combined with external factors/dynamics can "undermine" the system, cause a major break and lead to the establishment of a new agrarian system.

In the complexity and diversity of these historical processes, comparative agriculture endeavours to show the link between the remnants of former agrarian systems and the elements of the new appearing structures: locating continuities and permanences, discontinuities and ruptures, in short building a scaling likely to make the evolution of the agricultural sector of a society intelligible.

Making of the notion of agricultural or agrarian "crisis" an object of knowledge is pregnant with meaning: it leads to recognising that "agricultural development" in no way represents a "long quiet river" or a river only punctuated with rapids provoked by more asserted phases of technical progress. It leads to asserting the contradictory and differentiated nature of development and to measuring the ruptures and continuities making it up. "Taking crises seriously and making of them an object of knowledge, is to demolish the premise of the universality and indifference of the so-called economic laws, in respect of forms of social control which are unending sources of conflict between human groups; it is to assert that social systems have a history and that this history is meaningful, i.e. it accounts not only for the methods of social cohesion in current societies, but also for their relativity and fragility" (Aglietta 1981).

By taking into account the historicity of agricultural development processes, by calling on other social sciences (history, geography, economics and technological anthropology among others), by being concerned about identifying and characterising, for each agrarian system, what strongly resembles *a mode of accumulation* on the one hand, and *a mode of regulation* on the other, and by forming a theory thanks to the accumulation of studies *localised* in space and time, comparative agriculture does present a few analogies with the development of the "regulationist" school of thoughts in Economics.

4.4 Understanding Change in the Long Term to Better Identify the Process to Be Promoted

Beyond the type of research brought up previously and concerned with long term processes of crisis and change, historical research in comparative agriculture also looks at certain contemporary situations in a different light, and identifies problem areas that hinder the more "sustainable" development of agriculture today.

In this case, the diachronic approach proves to be suddenly operational and the research more *finalised* without appearing to. In the case of Burundi, as evoked earlier (Box 4.1 and Sect. 4.2), the detailed analysis of the agricultural revolution of the 1950s to the 1980s, looks in a new light at the agricultural policies managed in that country and at the many development projects inspired by these. Everywhere the idea was to encourage farmers to give up their complex intercropping to the benefit of sole cropping, even if it was the result of a remarkable process of agro-ecological intensification that was particularly simple as regards chemical inputs, and that relied almost exclusively on biological processes and meticulous work. Yet, while this intensification method had proved its worth, it was highly denounced by agronomists – dogmatic followers of sole cropping. With a strict assessment of these policies and corresponding projects, it became possible to rethink the future agricultural policy of that country on different bases (Cochet 2001).

Post-apartheid South Africa offers another example of how much the establishment, or reform, of agricultural policies cannot be conducted without in-depth knowledge of the historical processes leading to the situation of the time. In the former Bantustan of Ciskei for example which, as a result of apartheid policies, is very densely populated and covered with large villages, the first impression one had after observing this region at the end of the 1990s, was that of a landscape that was hardly cultivated, largely abandoned to the bush and apparently not very artificialised. One could observe more specifically forms of livestock production from grazinglands, many signs of erosion despite the spectacular recovery of woody plants, and a few occurrences of agriculture here and there. This agriculture, or what remains of it, is difficult to understand in that it results from a particularly contested history, where the more or less forced movements of populations in all directions were particularly frequent. While researchers, academics and development agents were wondering about the region's future and about the conditions of a new boost for "black" agriculture, we needed to stress the importance of a study aimed at identifying and understanding the former mechanisms that led agricultural activities to be abandoned (Cochet 1998). Indeed, in the 1950s and the 1960s, the region benefited from an apparently complex agrarian system, very much "alive" despite the disruptions suffered, and where livestock farming and agriculture were strongly associated. The discussions held with elders from the village of Twecu, at the very same place where they used to live before being grouped together by force, made it possible to piece together a few elements of this agrarian system: tillage by animal traction with intensive rotations and two harvests per year on the most favourable plots, animal fertiliser transported by cart towards the fields. It was the same for the gardens situated close to dwelling places, with fields enclosed by quickset hedges and/or embankments and low stone walls, and where a great many legume species were cultivated (idem). How did this agrarian system disappear, to the extent that, today, scrubland dominates the landscape, while rural population densities are higher than ever? How to direct the research of those who, today, would like to see this agriculture being "revitalised"? Piecing together the agrarian history of this region had become most urgent.

The policies being managed concerning the fight against slash-and-burn agriculture, in many countries of the intertropical zone, sometimes have quite the opposite effects of what was expected, precisely because this type of mode of exploitation of the environment is very often judged – and disqualified – without having really been studied or understood in the long run. Olivier Ducourtieux brings to light the fact that public authorities are determined to eradicate this form of exploitation of the environment, a determination based on the serious lack of comprehension of farming practices and their logics, and on a small number of simplistic ideas developed today as was the case in the colonial era. He describes the sometimes catastrophic effects of these policies, particularly in Northern Laos: acceleration of rotation and increased degradation of ecosystems, impoverishment of populations and forced movements (Ducourtieux 2010).

This call to history does not reflect a conception that would simply make of history "one more tool in the kit of the development expert".[7] The historical approach places the practitioner's intervention in the process in which he is trying to intervene. One must continuously "consolidate the idea that no one starts from scratch, that there is no blank slate; economic decision makers are never to be found at the beginning, but always in the middle of a series of events, each ensuing from all previous ones".[8]

[7] This expression is from Philippe Couty in: "Le temps, l'histoire and le planificateur", 1981.

[8] Idem.

Chapter 5
Comparing Productive Processes

5.1 Differences in Productivity on a Worldwide Scale and Consequences on Development

The growing globalisation of trade and the fact that farmers are increasingly being made to compete directly, makes the compared approach to production processes, productivities and incomes, on a worldwide scale, more than ever necessary.

On this point, we will recall that Dumont developed, in his research after the war, the first detailed comparisons of the gross productivity of agricultures, expressed in kilogrammes of cereals per working day.[1] Dumont was already guessing how much, showing these differences, was going to prove essential in understanding and anticipating the future evolutions of agriculture on a worldwide scale.

After Dumont and thanks to the compared approach to agrarian systems, Mazoyer then shed new light on the underlying causes of the crisis experienced by many farmers today around the world. Concerning the end of the nineteenth and the beginning of the twentieth centuries, and despite the extraordinary diversity of situations across the world, Mazoyer pointed out that differences in productivity among the different agrarian systems of the time were still very moderate. Between the manual subsistence agricultures that prevailed extensively and those that were already well equipped, thanks to the mechanisation of animal-drawn cultivation in Western Europe and the United States, net productivity, valued as cereal equivalent per labourer, was being established with a ratio of 1–10 at the most. These differences, in terms of net productivity per labourer, increased quite significantly thereafter.

In North-West Europe and the United States, motorisation followed by the rapid increase in horsepower, the significant increase in quantities of synthetic products being used (mineral fertiliser and treatment products), the adaptation of crops to these new means of production via genetic improvement, have led, since the end of

[1]Comparisons retranscribed particularly in *Economie Agricole dans le monde* (1954).

H. Cochet, *Comparative Agriculture*, DOI 10.1007/978-94-017-9828-0_5

the Second World War, to an unprecedented increase in yield, and more so labour productivity (Mazoyer 1987, 1989). Mazoyer showed, from the very beginning of the 1980s, that compared to productivity levels reached with arable crops in the cereal basins of these countries, cereal that was grown manually or with animal-drawn cultivation, as in most developing countries, was characterised by a labour productivity 500 times smaller which, in the context of the fast liberalisation of the time, was definitely leading to a crisis. The structural adjustment was in progress, imposing the opening of the borders to products of the North and the drastic reduction of public intervention, in the price domain in particular. Mazoyer also insisted on the significant consequences of this process of exclusion of the least equipped farming communities, on the worldwide explosion of unemployment, on the level of basic salaries in developing countries and on the emergence of a significant non-creditworthy demand in these countries.

Far from refuting Mazoyer's analyses formulated in the 1980s, the agricultural situation around the world during the first decade of the twenty-first century still confirms these tendencies. Indeed, Sophie Devienne showed that the best equipped farmers of the American corn belt today can cultivate 450 ha of maize-soya per labourer, thanks to very powerful tractors, sowers and very large combine harvesters, and by simplifying cropping techniques thanks to the use of genetically modified varieties. With yields of 100 quintals per hectare for maize and 35 for soya, a labourer can on his own produce 22,500 quintals of maize and close to 8,000 quintals of soya per year (Devienne 2002).

Faced with these extremely high levels of production per labourer, farmers of the Sahelian area, who have to deal with some of the most unfavourable soil and climate conditions in the world, cannot expect to produce the minimum quantity of grains required to feed their families. In the north of Burkina Faso for example, farmers equipped only with hoes and weeding instruments such as ilers,[2] cannot cultivate more than 0.6 or 0.7 ha per labourer, considering the short rainy season (4 months) and the urgent need to sow rapidly in order to avoid delaying the agricultural cycle (Guillaud 1993). With extremely low yields rarely exceeding 3 quintals per hectare in poor soils, an agricultural labourer cannot produce more than 200 kg of cereals per year of cultivation, i.e. barely enough to feed himself. Once wheat lofts are empty after several months, selling cattle enables labourers to gather enough earnings to buy provisions and wait for the next harvest.

In a family production unit of the high plateaux of Northern Ethiopia, the yields obtained (minus the seeds for the next cycle) are in the region of 2–6 quintals per hectare for cereals such as wheat or barley, 3–4 quintals for teff, the Ethiopian cereal, and 2–4 quintals per hectare for pulses such as chickpeas or lentils. For a farmer who is relatively well equipped with a complete set, i.e. an ard and a yoke with a pair of oxen, the maximum surface area that can be cultivated each year does not exceed 6 ha. By taking into consideration the fact that a farmer establishes crop

[2]Tool pushed in front of the labourer, with a long handle and fitted with a metal blade in the shape of a crescent to cut the roots of weeds in the shallow soil.

rotations, with the cultivation of wheat and barley in equal proportions on the one hand, of teff on the other and finally pulses, the production obtained (free of seeds) does not exceed, under the best conditions, a dozen quintals of wheat or barley, 8 quintals of teff and as much of chickpeas or lentils, i.e. barely the equivalent of three tons of cereals per labourer and per year (Cochet 2005). For those who have a complete equipment and enough land, animal traction leads to noticeable progress in productivity, progress that would be greater still under conditions that were less restrictive than those of the plateaux of Northern Ethiopia.

Gross production differences per labourers would then be even more significant today: American corn belt farmers would be in a position to produce 10,000 times more food than their African colleagues situated in the worst conditions and having only manual tools at their disposal, i.e. 1,000 times more than an African farmer with a yoke still.

Of course, this is not about productivity, for the establishment and functioning of these production systems with high-tech equipment and input consumption is expensive. But the difference in productivity separating African farmers from their North American colleagues is huge in any case, in the region of 1–500 for those with access to tillage by animal traction, and rather in the region of 1–2,000 or 3,000 for those working with manual tools.... And yet all these producers are competing directly on the same international market with prices dictated by the most productive farmers, and all the more depreciated by the State subsidies paid to the farmers of the North.

These figures,[3] as simplified as they are, also show the scale of issues when thinking in terms of development for a continent like Africa. How could African agriculture survive this inequality of opportunity, continue to feed cities at very low price where world cereal surpluses are dumped (sometimes in the form of food relief), become better equipped, reconquer parts of the market and develop? And how could it remain confined to crops less challenged by the countries of the North such as coffee or cocoa which, as they say, benefit from a "comparative advantage", when the evolution of their prices follows the same path and when substitutes start to take a significant place in food-processing plants (vegetable fat in the chocolate industry, artificial flavours, etc.)?

Developing countries were strongly encouraged, in the 1980s, to specialise in tropical productions needed by the North (coffee, tea, cocoa, tropical fruit, etc.), which means that not only they neglected their subsistence sectors, but these were also unable to withstand the competition from the countries of the North with their cereal productions. Today, investors from the countries of the North are coming to these developing countries to start vast farms specialised in the production of the very same grains (cereals and oil-producing plants) that had been eliminated by the Northern competition thanks to its comparative advantages.

---

[3] Such comparisons could be extended to other cultures, for example in the case of cotton for which differences in productivity are as significant between the cotton producing regions of the United States, and the cotton areas of West Africa (Dufumier 2004).

5.2 Which Forms of Agriculture Should Be Promoted?

Many pleas "for" so-called "family" agriculture have seen the light these last years, coming from activist groups (the via *campesina* international network for example), research institutes (CIRAD, in particular for the Paris International Agricultural Show in 2005), groups of countries within the framework of international negotiations (Europe, Japan, Switzerland, by highlighting the multifunctionality of family agriculture), or even from international organisations such as the World Bank which has been discovering its virtues, particularly as regards fighting poverty within the framework of the Millennium Development Goals, but also as regards its productive and job creation capacity.[4]

Moreover, family farms often ended up dominating throughout the second half of the twentieth century, by progressively replacing very large farms that were based on different social relations: regression of very large latifundia-type properties in the South of Europe and Latin America (agrarian reforms, division through inheritance, selling pieces of land), bankruptcy of colonial plantations and triumph of small planters (in Côte d'Ivoire, for example), dismantling co-operatives and State farms in many countries of the former Soviet Bloc or depending on it, among others. As to Western Europe, while Karl Marx was anticipating at the end of the nineteenth century, the planned disappearance of the "farming production method" and the advent of large capitalist farms with salaried employees, there again it was the family farm that laid down the law, including in the large sectors of North American agriculture.[5] The contemporary agricultural revolution and the tremendous productivity gains resulting from it were carried through to a successful conclusion by family farms, although the unprecedented increase in the level of working capital (and actual capital owned by farmers) did not make of these family farms "capitalist" farms, on the contrary.

From a strictly economic point of view, let us recall that among the reasons for its historical efficiency and its resilience, high on the list is the very specific nature of the Land/Capital/Labour relationship, in which, unlike the model proposed in his days by Marx, (1) capital does not yield an average interest rate, (2) labour, which is essentially family-based, is frequently under remunerated or less than the average hourly wage, and (3) the land, especially when it is solely owned by the family, is not remunerated at the average rate of the ground rent. The effect of the economic result of the production process for the producer was to obtain a farm income, a notion which is far removed from that of profit or rate of return on capital invested.

In many situations, family farming widely showed its efficiency, particularly (1) as regards wealth creation per unit area (net value added/ha) thanks to more intensive combinations of production factors, labour in particular, (2) as regards job

[4]In *Rapport sur le Development dans le Monde*, report analysed in Mazoyer et al. (2008).

[5]Where, unlike an idea which is firmly rooted in people's minds, the great majority of production units remained family-based, particularly in the sector of arable crops (Devienne et al. 2005).

creation and maintenance, as well as to ensure that the family workforce appealed to was fully employed throughout the crop year, with a diversity of tasks to be fulfilled within often complex and diversified production systems, (3) as regards labour productivity also, because the other forms of production so often showed low efficiency in this domain (latifundia, State farms, production co-operatives), (4) and, finally, as regards the management of complex agro-ecosystems thanks to farmers' knowledge of the environment and historically acquired know-how.

Today, environmental concerns together with the desire to have better quality and healthier food, reassert (e.g. in France) the virtues of family farming that would be rooted in a territory, supplying quality products, managing ecosystems, creating jobs and incomes, and taking part in the living rural fabric.

Although the significant increases in productivity recorded in the French agricultural sector after the war, are on the whole due to "family" farming, it would be interesting to know whether they did not reach a sort of limit that was difficult to surmount within the exclusive framework of "family" farming. The recent technological evolutions (robotic milking system to break down the barriers of milking in dairy farms, direct sowing to avoid certain labour peaks in general cropping farms, GPS and so-called "precision" farming to face the growing difficulties of managing increasingly larger and therefore increasingly heterogeneous plots, increase in the furrow width of machines and increased automation) show that pursuing productivity gains is still possible within the framework of the same family "model". But other innovations, this time organisational and social innovations (although these always happen together with capital increase), indicate that other potential productivity gains could be found elsewhere: "pooling" of herds within the framework of "dairy partnerships" to make the milking and discharge management facility profitable, new forms of associations, collective rotations managed through a farm machinery co-operative (CUMA) and through an employer group, agricultural production companies completely free from the permanent control (via a lease or title deed) on land,[6] among other things. These new institutional forms openly questioned the French-style "family" agricultural model, which was the foundation of the consensus established at the beginning of the 1960s between farmers and the public authorities.

Elsewhere in the world, the takeover of important agricultural areas in the countries of the South and the former Soviet Union by public or private foreign powers, a movement which occurred before but which took an unprecedented scale with the surge in agricultural prices in 2007–2008, was also questioning the future predominance of family farms in many regions of the world. While the governments of countries relying widely on the international market to stock up with food and/or agrofuel, decided to directly take care of their supplies *off shore* without going through the international market, private investors saw in it an opportunity to make a significant profit while diversifying their business portfolios.

[6]All these institutions being from now on recognised by the Framework Law on Agriculture of 2005–2006 (Cochet 2008a).

This large-scale movement of acquisition or control of agricultural lands by large and often foreign firms, is a new phenomenon because of its scale. Today, nothing seems to be able to stop it, nor can any one measure its consequences or even quantify its extent. The questions raised by these investment projects as regards political and social risks (non-transparency of land transactions, the terms and conditions of the local governance of land and water resources not being sufficiently taken into account, potential eviction of local populations), as regards food security (exporting basic agricultural products when the food security of local populations is not guaranteed, replacing food crops with crops intended for the production of agrofuels) and as regards environmental risks (development of cropping system favouring monoculture, the massive use of synthetic inputs and risk of soil and water pollution, reduction of biodiversity) gave rise to many reports being drawn up by certain NGOs as well as governments and international organisations. The issue of their economic efficiency also needs to be raised (*infra*, Chap. 11).

These evolutions which can be observed in dribs and drabs in many regions of the world, reveal a new capital/labour relationship, which is progressively and increasingly clearly moving away from the family farming situation: a progressive separation between capital and labour, where the holder of the capital no longer or decreasingly lends a hand, while the labourer brings in less and less capital to the production process, even if he is the owner of the land involved in the production process. Such evolutions would lead to a loss of autonomy, a partial or total loss of control of the production process for the farmer who is a party to contractual farming. Such evolutions would inexorably lead to the rapid proletarianisation of farming, thereby vindicating more than a century later what Marx (and others) had anticipated, even if the nature of the new social relations established today, differ from the ternary relation defined at the time (landowner, capitalist farmer and salaried farm labourer).

It is important to point out the existence of "counter-examples", for which the "liberalisation" and arrival in force of new actors (with capitals outside farming), did not in actual fact reflect this type of evolution. Contract farming developed in the lowlands of Tesechoacan (Mexico), offers an example where the private sector and the State intervened jointly in a process of capital modernisation and accumulation, in a type of farming that remained mainly family-based and modest in size (Brun 2008). At this stage, we urgently need to understand better the conditions to be met for these processes to become possible, and to actually lead to a reinforcement of family farming (its efficiency and viability), whereas the evolutions described previously lead rather to its weakening.

The future of these very different forms of agriculture, beyond the great diversity of forms which each concrete and regionalised form of agricultural development can take inside these two major groups, will depend mainly on the political choices made nationally and internationally. These should depend particularly on the performance criteria to be privileged: financial efficiency, economic efficiency, job and income creation capacity, lower fossil energy consumption, minimization of chemical pollutions.

Photographic credits
Plates 1 to 6 H. Cochet
Plate 7: H. Cochet and J.D. de Surgy
Plate 8. H. Cochet and O. Ducourtieux (bottom)

Plate 5.1 Agrarian landscapes in Burundi (African Great Lakes Region). (**a**) Agro-pastoral highlands in South Mugamba in Burundi, between 2,000 and 2,200 m above sea level: semi-scattered settlements found partly on the summits, with contiguous plots of cereals and associated pulses yielding two crops per year thanks to careful manuring, joint pastures in the lower slope sections, forest residues in thalweg heads, lowlands reserved for grazing during the dry season (photo H. Cochet, May 1993). (**b**) Grazinglands and associated transfers of fertility disappeared long ago, gradually eaten away by increasing cultivated areas and settlement densification. Dense banana plantations on the summits and field patterns on the slopes: small wheat parcels sometimes associated with banana trees on the upper slope section, complex food-producing associations on the lower slope sections, and a few plots of tea and coffee as seen in the centre of the photo (1,800–2,000 m above sea level) (Photo H. Cochet, July 1992)

Plate 5.2 Ukraine, dual farming. (**a**) Photo taken in Ukraine from the sky. On the left, we can see a street-village alongside a valley and a plateau with very small plots: these are enclosed garden-orchards around houses and plots situated along settlements, perpendicular to the main roads, the whole resulting in very small labour-intensive mixed crop-livestock farming operations of 0.5–1 ha. On the right, we can see very large fields, also inherited from the Soviet era and currently owned by private businesses, where large-scale motomechanised cultivation takes place over several thousands of hectares. On the top right, we can see "reserve" plots allocated to certain villagers, leading to the widening (2 ha) of "people's" farms. (**b**) Photo in Ukraine at ground level. In the foreground, we can see the houses of a village alongside a small valley, with enclosed garden-orchards and forage cereal plots, potatoes and vegetables growing along houses. In the background, we can see very large fields with annual crops (Photo H. Cochet, June 2010)

Plate 5.3 Agrarian landscapes in Bolivia. (**a**) Potato field under preparation (ploughing) in a cropping pattern regulated according to the community (*aynoka*). Potato crop comes in as rotation head in cultivation cycles that still give more than their due to multiannual and grazed herbaceous fallow land, for soil fertility regeneration (3,200–3,600 m above sea level). Settlements are half-scattered on the projecting ledge at the bottom of the slope and overlooking the water stream (Altamachi, Cochabamba Mountain Range) (Photo H. Cochet, June 2003). (**b**) Agricultural colonisation patterns in the eastern plains of the Bolivian *Oriente*, East of Santa Cruz. At the top, the keyboard- and star-like shapes correspond to colonisation projects planned by the authorities to provide migrants from the Andes Cordillera and in search of land with plots. At the centre and to the right, we can see large agribusiness farms (of several thousands of hectares of extensive livestock farming or soya/sunflower farming); at the bottom, we can see old Mennonite colonies

a

Plate 5.4 Agrarian structures and mode of exploitation of agro-system in the Ecuadorian Sierra. (**a**) At the bottom of the valleys, the best lands are always part of large farms. The agrarian reform redistributed the surrounding land to former labourers, but the centre of former haciendas, at the bottom of irrigated valleys, were kept by the former owners running large motomechanised farms specialised in dairy farming. On the slopes, we can see small farms practicing mixed crop-livestock farming (Photo H. Cochet, May 2003). (**b**) Fragmented fields can be seen here in the Andean valleys at around 3,000 m above sea level, with cereals, pulses, quinoa and market gardening. In the foreground, we can see strips of plots running in the direction of the slope, as a result of divisions in property inheritance. Rare eucalyptus trees can be seen here and there near the settlements in particular (Photo H. Cochet, May 2006)

Plate 5.5 Ensete (false banana) Gardens, cereal growing and associated livestock farming in Southern Ethiopia. (**a**) Photo taken in Southern Ethiopia, the "green" part of the country. Scattered around the settlements are ensete gardens, vegetable gardens and small coffee plantations. Beyond that, we can see maresha-tilled plots sowed and rotated with cereals and pulses, as well as pastures and associated livestock breeding, supplying draught power and manure. Trees are omnipresent, particularly eucalyptus for firewood and timber (Photo H. Cochet, May 2000). (**b**) Photo taken in the Gurage zone where ensete gardens serve as basic food-producing sites for rural families, and occupy all the upper slopes near the dwellings. Plantations are carefully fertilised thanks to the meticulous recycling of dung from herds walking through the settlements at night (Photo H. Cochet, June 1998)

Plate 5.6 Destroyed forests, built forests. (**a**) Extensive livestock farming frontier in Uxpanapa (Mexico): deforestation, introduction of temporary grasslands and hasty installation of barbed wire enclosures, in a context of share livestock farming (Photo H. Cochet, May 2001). (**b**) Aerial photograph of a coffee agroforest system in South-West Ethiopia. A few decades ago, the land on both sides of the road was completely deforested; it was tilled with a maresha and sowed with cereals and pulse plants. Since then, a diversified tree cover has been completely reconstituted, sheltering the arabica coffee plantations

Plate 5.7 Landscapes in Eastern France (Lorraine). (**a**) In Eastern France, priority is given to equipment for large-scale farming. The contemporary agricultural revolution contributed to shaping, simplifying and homogenising landscapes for greater labour productivity. On this photograph, rotational cropping concerns small grains and colza as rotation head (Photo J. D. de Surgy, May 1995). (**b**) Non-consolidated plots towards Sarrebourg. For a long time pluriactivity remained the rule in the vicinity of the coal basins, resulting in different agricultural landscapes. In the background, the Vosges (Photo J. D. de Surgy, May 1995)

Plate 5.8 Landscape builders. (**a**) Irrigated lucerne fields in a valley on the Pacific side of the Central Andes in Peru (Cotahuasi, Arequipa). Inclined terraces on the colluvium at the bottom of small valleys, with gravity irrigation and dairy farming. The plantation of eucalyptus ensures the supply of timber and firewood (Photo H. Cochet, September 1983). (**b**) Rice-growing in the mountains of Danian (Province of Guangxi, South China). Meticulous planning of terraces, monitoring of run-off and managing irrigation individually. At the bottom, we can see steeper slopes covered in and planted with trees (Photo O. Ducourtieux, May 2009)

Part II
Methods and Expertise of Comparative Agriculture

Where theoretical knowledge on comparative agriculture was built *from the bottom*, comparative agriculture was born at first as an approach to agriculture, a long term practice, a "know-what-to-see" and expertise before becoming a scientific approach. The progressive generalisation of its basic notions, tools and concepts results from a long experience in the field, and from a perspective on its main results over increasingly diversified historical periods and geographical areas.

This expertise and these methods are the subject of this second section. They made it possible to mark out the path to follow to carry out the diagnostic analysis of an agricultural region,[1] an approach implemented to identify ongoing changes in the said region and the development paths of the different production systems in a perspective of regional action (project identification or impact evaluation for example). They also led to marking out the path of the evaluation per se, i.e. the systemic evaluation of the impact of a project or measure of agricultural policy, or the economic evaluation of an agricultural development project from the general interest point of view.

This expertise relies first of all on the choice of the agricultural region as privileged scale of agrarian dynamics understanding, and on how carefully the landscape is being read. It also relies on a field work approach giving an important place to surveys conducted with farmers in particular, and aiming at collecting the qualitative elements required for understanding processes and, at the same time, the rigorously quantified elements required for measuring these processes and their

[1]This approach of the diagnostic analysis of agrarian systems was first taught at the Institut National Agronomique of Paris-Grignon (today AgroParisTech) by the Comparative agriculture and Agricultural Development Unit. It was then hived off towards other higher education and training institutions, in France (IRC in Montpellier SupAgro, ISTOM) and overseas (Mexico, Brazil, Ecuador, Senegal, Vietnam and Laos among others), as well as towards international organisations, the FAO in particular. Several articles and teaching books have been dedicated to it, particularly Devienne (1997), Devienne and Wybrecht (2002), Apollin and Eberhart (1999), Dufumier and Bergeret (2002), Cochet et al. (2002), Ferraton et al. (2003), Cochet and Devienne (2004, 2006), Ferraton and Touzard (2009), and the FAO (1999).

results. The development of a specific expertise, particularly to collect, through direct surveys, the elements required for reconstructing former and contemporary dynamics will then be analysed. We will also take an interest in the way farm typologies developed in comparative agriculture are built, and in the necessary prior identification of production systems in particular. In a fifth paragraph, we will examine closely the economic dimension of analysis using comparative agriculture criteria, the careful quantification of value added and productivity on the one hand, its distribution and income formation on the other. The issue of evaluation will then be tackled. The expertise developed in comparative agriculture proved to be very efficient indeed in identifying and measuring, as close as possible to reality, the differentiated impact of development projects, programmes and policies on producers and agrarian systems.

Chapter 6
Micro-Regional Approach to Agrarian Issues

6.1 Agricultural District: Privileged Object of Analysis in Terms of Agrarian System

The regional approach was "invented" long ago already by geographers. Referring only to the French school of geography, we should mention Paul Vidal de la Blache, its founder (as early as the end of the nineteenth century) and, after him, Pierre Gourou, Gilles Sautter, Paul Pélissier, Jean Pierre Raison for the tropical areas of Africa or Asia, and Olivier Dollfus for Latin America, among other researchers. Others followed, choosing to work on a wider scale – the *terroir*, the village territory – or, on the contrary, at the national level.

The regional approach is also part of the comparative agriculture approach. But what would be the ideal dimension of the region under study? This question refers to the issue of scales and frontiers mentioned previously, and to the necessary conjoining/combination of scales of observation and analysis (*supra*).

The "agricultural district" was established as a privileged level of analysis and understanding of the agricultural activity more than 50 years ago. In 1954, Malassis, Cépède, Bergmann, Dumont and others published *La petite région agricole, contribution à l'étude et à la réorientation de l'économie agricole d'une petite région* (The Agricultural District, Contribution to the Study and Reorientation of the Agricultural Economy of a Small Region), which aimed at devising a well thought-out and coherent method of study prior to the modernisation of agricultural regions. The approach in terms of agrarian systems was not yet on the agenda, but the choice of agricultural district, as a way in for the analysis, was perhaps foreshadowing its advent. This was like suggesting that farms were rooted in a *territory*, that they partook of a *rural fabric*, and that the conditions of the environment, the different *ecosystems*[1] to which they had access, constituted the working tool of farmers as much as their living environment.

[1] Words that were not used in those days . . .

H. Cochet, *Comparative Agriculture*, DOI 10.1007/978-94-017-9828-0_6

What must we understand by "agricultural district"? Probably a district with a minimum dimension leading to the perception of the entire agrarian system (even if it covers a much vaster geographic area), i.e. practice/ecosystem relations founding one or several modes of exploitation of the specific environment, differentiation mechanisms internal to the system, social relations and regulation modes coherent with this mode of exploitation of the environment. The actual process of understanding agricultures, carried out at the level of the district, leads to an in-depth understanding of the mechanisms and processes at play. It then becomes possible to widen the perspective, as long as the districts under study have been chosen judiciously and lead to addressing the issue of agricultural policies at the regional or national scale.

Choosing to tackle the complexity of agricultural issues on this scale of analysis and to develop a systemic approach, necessarily leads to the study of all forms of agriculture[2] developed on that territory and not only of one specific production or another, or one category of producers or another, as will be seen as regards typologies (*infra*).

6.2 Reading the Landscape

In comparative agriculture, landscape is the visual expression, what can be seen at a given scale of observation and *mode of exploitation of the environment*, itself a party to an *agrarian system* (*supra*). The works of Deffontaines, in the French context, showed how much an analysis of the landscape could prove fruitful for the study of the agriculture of a region and the production systems: "The agricultural production systems of a region are partly in line with space; moreover, the landscape can be perceived as the support of original information on many variables, relating to production systems in particular, with their superimposition or proximity revealing or suggesting interactions" (Deffontaines 1973). Observing the landscape revealed practices; what could be seen suggested what was functional (Deffontaines 1997).

In comparative agriculture, deciphering a landscape, based on its detailed and methodical observation, consists in delimiting its different parts so as to describe each one better,[3] and in deducing from what can be seen of the customs and practices, at a given time, a number of hypotheses on the mode(s) of exploitation of each one of these parts, as well as on the possible relations between the different exploited spaces. It goes without saying that the observations and deductions/hypotheses emerging from such a reading, are in line with different scales of analysis. What can be seen of the cropping practices refers rather to the scale of the *cropping*

[2]The word "agriculture" being taken in the wider sense here, and which of course includes livestock farming activities and, if need be, fruit gathering, fishing and forest-related activities.

[3]Gilles Sautter's and Chantal Blanc-Pamard's landscape "facets" for example (Blanc-Pamard 1986).

system, while the major mode(s) of exploitation of the environment express their coherence on the encompassing scale of the *agrarian system*. The fact remains that the relations suggested by, for example, the juxtaposition of several easily "readable" *terroirs* in the landscape (meadows in humid valley bottoms, cultivated slopes, planted south-facing hillsides, dry shrub steppes grazed by a few sheep herds, to only give the simplest examples) inform just as much about the conceivable combinations of different cropping systems and livestock systems, this time at the level of the production system, as about the possible access to different resources and the evolution of their exploitation (presence of fallow lands, for example). "We start from what is visible to then decipher the systems contained in and acting on space" (Deffontaines 1973).[4]

Some geographers had already made of the study of landscape a privileged tool of comprehension of spaces and human activities, and very early had highlighted (Pierre Gourou) that a landscape was not self-evident, that its analysis required to "be careful" and to apply rigour. For Chantal Blanc-Pamard and Pierre Milleville, "the analysis of the agrarian system goes through a careful reading of the landscape that appears as a farming construction, as the result of agricultural practices based on the farming perception of the environment. It is like having a dialectic relationship between practices leading to a landscape, and the landscape as an expression of practices" (Blanc-Pamard and Milleville 1985).

Observing practices and reading the landscape go hand in hand in this approach to the terrain, which comparative agriculture shares with certain ruralist geographers. To make farmers talk in front of and about "their" landscape is also, as will be seen subsequently concerning field surveys, an opportunity to give a voice to those who make of the landscape their tool, to those who have fashioned it, built it or, on the contrary, deteriorated it.

As the first indispensable step in identifying and analysing the mode of exploitation of the environment characteristic of a given agrarian system (*supra*), reading the landscape is a delicate and demanding exercise.[5] Moreover, observation, as meticulous as it is, is not sufficient in formulating hypotheses on the farming method if the result of this reading is not actually organised, categorised and finally modelled thanks to one or several characteristic transects or block diagrams.

Today, and although the growing artificialisation of environments permitted by contemporary agricultural revolution, on the one hand, and the total amount and terms of the distribution of subsidies to farmers depending on the successive reforms of the PAC, on the other, have led one to believe for a time that farmers' decisions could partly free themselves from the conditions of the environment, the

---

[4]Concerning the methods of observation of the landscape and their utilisation as regards understanding farming practices and production systems, we refer the reader to the works of INRA-SAD on the Vosges Mountains (INRA 1977), or Alpine for the French context or, for developing countries, to those of Gilles Sautter (1985) or Chantal Blanc-Pamard (1990). See also the educational work of Lizet and de Ravignan (1987).

[5]Today *Google Earth* is a remarkable and particularly valuable tool for reading and analysing a landscape "from the sky" in high resolution, during fieldwork.

landscape analysis and its detailed "reading" remain inescapable in apprehending the study of the production systems of a region. Furthermore, the return in force of the notion of *territory* – though it would be advisable to define this term which sometimes does have very different connotations – in environmental preoccupations, encourages more than ever agroeconomists and geographers to meet halfway around the landscape and, beyond, around the regional space.

Chapter 7
Terrain and Surveys

7.1 Conducting Surveys Oneself and Gathering Qualitative as well as Rigorously Quantified Data

Although today, thanks to the many studies conducted by researchers and practitioners from different disciplines, we know more about agricultures in many regions of the world as well as their histories than ever before, their contemporary transformations, irrespective of their characteristics, will still be better apprehended and understood if they are the subject of studies conducted in the field. This field work requirement remains more than ever a basic principle of any research or approach to comparative agriculture, and not only a compulsory passage for a young researcher at the beginning of his career. It is not either about mere field visits, like those experts in too much of a hurry or those officials and elected representatives looking to renew their mandate who "drop in on the field", but indeed about spending long periods of time in a given region, and patiently carrying out observation and listening.

Whether the field work approach is carried out personally or with a team, it could not be delegated to just about anyone, unless such delegation was carried out within a careful learning process. Indeed, data gathering, processing and interpreting remain indissociable; that is why an interviewer and a researcher *necessarily* merge into the same person. Observing (landscapes and practices), questioning and listening must be organised as one goes along through a system of hypotheses which is itself built around clear concepts. If the researcher must also be the interviewer, it is also because the choice of interlocutors – and therefore the sample of farms under study – is built up step by step, based on the prior reading of the landscape and a historical approach, as will be seen as regards typology construction (*infra*).

An observation and survey *method* specific to this discipline was developed by the Comparative agriculture Department of AgroParisTech. While it contained all the necessary rigour, it did away with the sometimes impressionistic feel of

H. Cochet, *Comparative Agriculture*, DOI 10.1007/978-94-017-9828-0_7

Dumont's travels. Indeed, it did not consist in elaborating thick questionnaires entrusted to many interviewers scattered in the field, including wide points of departure on statistically "representative" samples with results processed by powerful software programmes. This method sought more to reveal cause and effect relations that could explain ongoing processes, than searching for statistical correlations between observed phenomena.

Our "way of doing things" as regards surveys in rural environments, is closer to the ethnographic-type survey in that it involves observation, listening, immersion in the environment, daily note-taking and absence of closed questionnaires, as well as through its duration, necessary contextualisation and non-delegation. However, it differs slightly from the ethnographic-type survey in that it requires a *judgment* sampling and is concerned with *quantification*. By quantification, we must understand ultimately obtaining quantified and comparable results (productions, costs, quantity of work, value added, incomes... *infra*), and not systematically resorting to quantitative methods of data processing,[1] which incidentally are not used very much in comparative agriculture. While the greatest care is given to the choice of questions, their linking, the simultaneous (and sometimes participating) observation of the objects, tools, machines, skills and their sequences, it is this position of the researcher-interviewer that can best lead to obtaining the most reliable information possible, the basis of any comprehensive quantification of ongoing phenomena.

Furthermore, mastering concepts of comparative agriculture (see Section One) during interviews conducted with farmers, leads to asking the right questions, to find the pieces of the puzzle that were missing, to search for elements that must *necessarily* be present for reality to make sense, in short, for the *system* to exist and to be able to *function*. For an element collected by the survey to become a *fact*, to make sense and finally to be interpreted and, in this respect, to find its rightful place on the puzzle, it must be understood through concepts that will give it its place in the theoretical construction underlain by this exercise. As written already by Couty in the 1980s, the idea is to "link facts so as to set them up in a meaning-creating sequence" (Couty 1984).

Concerning surveys on cultivation practices, Sébillotte had already pointed out that "transforming observed phenomena into interpreted results supposes the construction of a theoretical model which in turn validates one reading or another of reality" and that "it is indeed through a dialectic play between the theory that guides the survey, and the organised observation that leads to controlling the questions and answers, that we will be able to avoid the bulk of the many traps threatening any hasty interpretation" (Sébillotte 1989).

This is where we can measure how much the opposition between "empirical" and "theoretical" research, which is still very much alive or implicitly suggested in the scientific community, is removed from comparative agriculture. A scientific

[1] "Quantitative" data processing methods are called like this in reference to the statistical tools being used which involve a large sampling, and not in reference to the intrinsic nature of the (qualitative or quantitative) data collected.

"division of labour" between, on the one hand, those who go on site and collect data and, on the other, the theoreticians who supposedly have a monopoly on conceptualisation – a division which is almost generalised in many disciplines and research institutions – could have very serious implications not only because of the risk inherent to the separation of tasks for the reliability of the data processed, but also because of its underlying relations of domination. Where empiricism consists in "going on site", i.e. in multiplying and classifying *localised* observations, it means that the field work approach structures this discipline as much as it justifies its identity. But when these *comprehensive* descriptions of reality which are prompt in handling the systemic approach and change in scale, and when these observations and analyses, rigorously conducted within a coherent conceptual framework, contribute to elaborating the theoretical part, then we can measure how much this empirical/theoretical dichotomy appears largely nonsensical.

7.2 Knowledge from the Top and Knowledge from the Bottom: Breaking Implicit Hierarchies

When interlocutors are not exclusively chosen from among "model" or "technified" farmers or those "who go forward" – all these expressions generally bearing value judgments, observing and listening to these interlocutors often reveals practices that were ignored, usually misunderstood and sometimes scorned. They account for visions that are more varied without appearing to; they shed light on the complexity of practices and reveal their meaning. Concerning ethnographic surveys which, through certain aspects (*supra*), are close to surveys in comparative agriculture, Stéphane Beaud and Florence Weber wrote: "The ethnographic survey in contemporary societies is not a neutral tool of social science. It is also the instrument of a struggle which is *scientific and political* at the same time. The ethnographer is by definition the one who is not satisfied with overhanging visions or with already existing categories of social descriptions (statistical categories, categories of prevailing or standardised thought). He expresses scepticism in principle as far as generalist analyses and pre-established social divisions are concerned. The ethnographer reserves the right to doubt *a priori* the ready-made explanations of the social order. He always cares about coming closer to social reality, even if it means going against official visions, opposing the powers that impose respect and silence and that monopolise how the world is viewed" (Beaud and Weber 2003).

Comparative agriculture shares this vision of the survey. Only field work can truly give their voice back to those who never speak and break the implicit hierarchies that come too often between "scientific" knowledge and "popular" knowledge, between those who know and those who do, between agronomists and farmers. Because agricultural practices are extraordinarily diverse across the world, because farmers, in their immense majority, kept away from the spheres of power and as such were kept away from the elaboration of policies and projects that concerned them, agriculture is perhaps the domain of activity where this field requirement is the greatest, and where the knowledge potential it reveals is the most important.

In the French context, with the many regional agroeconomic studies conducted jointly by lecturing researchers and students from the Comparative agriculture Department, we can assert that in all the regions studied, interviews with farmers made it possible to catch a glimpse of what a dialogue cleared of "developmentalist" agendas and value judgments could reveal, particularly when it was *also* conducted with those who were "excluded" from "development", i.e. all the farmers who are rarely recommended by agricultural or "extension" organisations, those who were still found in large numbers in the 1980s and the 1990s and whose turnover was too little to be charged on the real basis, and who were therefore absent in most files and statistical databases (RICA, in particular[2]). For example, the dynamics of farm extension and their reorganisation, which have been analysed and described so many times, even though they have been a major phenomenon in the last 50 years, cannot really be apprehended as a whole without analysing, in detail, the mechanisms underlying the progressive marginalisation and disappearance of the least "successful" farms, a phenomenon which is no less major since it leads to the extension of neighbouring structures. Moreover, and to stay in the French context, it is often with this type of interlocutors that one can discover so many practices, skills or so much knowledge which are called upon today to assist "sustainable development", the multifunctionality of agriculture or the agroecological transition.

This finding is even more striking in many countries of the South or "developing" countries. Save for rare exceptions, including those of Dumont and de Schlippé, we can say that, during the entire colonial period, the study of practices was left to ethnologists or "folklorists" (Sigaut 2004). When agronomists "made it" to the field, whip in hand, it was most often to establish "special crops" that were needed back in the home country, or to develop food crops "in the interest of the indigenous communities" with imported normative crop management sequences, and in the name of the necessity to "put the farming community to work". I have already shown how, in a country like Burundi, the colonial agronomists' ignorance of local farming practices led the populations to sink dramatically into starvation, and to a sudden stop in the labour intensification process that had yet seen the light (Cochet 2003). After the Independences, most national agronomists as well as their teachers and European "counterparts", did not look into farming practices which had been rejected as part of the "traditional" or "archaic" domain, after being measured against the yardstick of development models then taught in the West as well as the East. How many times have livestock farming systems been described as "contemplative" so that researchers would not have to be confronted with reality, and cropping systems been likened to "gathering" because they were indiscernible through the normative filters of standard agronomy? Several decades had to go by

[2]RICA: Réseau d'Informations Comptables Agricoles (Agricultural Accounting Information Network).

before the local "know-how" became a real object of detailed studies, although too rarely, and not always by agronomists. In 1976, when agronomists were still too few to look into farming practices without prejudice, Sigaut could still write: "It is not by chance if we owe ethnographers the essentials of what we know on the former and exotic agricultural techniques" (Sigaut 1976a).

That is why one of the main issues in comparative agriculture, as far as training in higher education is concerned, is to ensure that students spend time "in the field". Beyond the corpus of knowledge they can acquire and the new way of accessing knowledge field work represents, submitting to this field requirement is also a privileged way of discovering a new know-how and, more still, a new know-how which is essential for any development practitioner in the North as well as in the South, a sort of *scientific humility* that undoubtedly would have prevented many failures as regards "development".

7.3 Tools, Machines and Skills: Agricultural Technology and Innovation Problematics

Placing field work and interviews at the centre of the methods developed in comparative agriculture, gives the greatest importance to the observation of tools, machines, skills and positions, to the in-depth study of practices, and therefore to the comprehension of the implementation of knowledge (from the bottom and the top). Each tool and each instrument must give rise to a triple deciphering: a description of the tool, a description and analysis of the functions of the tool (linked to the analysis of the environment and the technical sequence or crop management sequence), as well as an observation and analysis of the way the tool is used (actual gestures and their sequence, machine tuning and driving). A similar approach can be developed for farm inputs.

Agricultural technology or technical anthropology offers an important support in this regard. In France, the works of André G. Haudricourt and Mariel J.-Brunhes Delamare (1955), Haudricourt (1987), and those of François Sigaut (1975) for example, have made decisive contributions to the comprehension of technical evolutions in agriculture, thanks in particular to their comparatist approach. Although today this scientific discipline is not very developed, with the so-called "cultural" approaches having sometimes pushed the study of tools and skills into the background, more recent works have been carried out on African ploughing tools and put together into collective works at Christian Seignobos's instigation (1984, 2000). Several scientific journals are still specialised on the issue: *Techniques and Culture* (CNRS/MSH[3]), *Tools and Tillage* (International Secretariat for Research on

[3]French National Centre for Scientific Research, Human Sciences Centre.

the History of Agricultural Implements), and *Journal d'Agriculture Tropicale et de Botanique Appliquée JATBA* (MNHN[4]) to a lesser extent, among others.

By comparing tools and techniques at different periods and in different places, these technologists have established technological *relations* as well as progressive dynamics, and in that have shown better than anyone else the absurd nature of the qualifier "traditional", which is understood as immovable or unchanging, and is wrongly assigned to one technique or another. Sigaut has shown that, for example, by specifically analysing the *functions* of the ard (burying seeds well before ploughing) in different agrarian societies, we could catch a glimpse of one of the reasons for which, historically, there is almost no animal traction in Sub-Saharan Africa, while those who unsuccessfully popularised its usage invoked the "cultural" refusal and "resistance to innovation" of the populations in question. "The problem is not to try one 'innovation' or another blindly, like trying all the keys of a bunch one after the other to find the right one. The problem is, first of all, to understand the lock, i.e. the internal logic of the existing cropping systems. This will require in-depth analyses, based on the observation of actual practices which is detailed, precise and as objective as possible, and on the very large comparison of many observations" (Sigaut 1976a: 17).

By carefully analysing technical and scientific innovations, Sigaut also showed that technical changes had very often preceded the development of scientific knowledge, which reversed the relation between science and technique in a peculiar way (idem, p. 20). The issue of agricultural innovation, which is so important as far as agricultural development is concerned, is raised in a completely different manner.[5] Perceiving and bringing to light all the potential of "endogenous" technical innovations in the so-called "traditional" practices of farmers, is to put back in their context, differently and with more chances of succeeding, innovations from elsewhere and the possible transfer of knowledge from one society to the other, or from one group of farmers to the other. When sociologists and psychologists are called upon to try to understand why the "technical message" is not going through or well with farmers, two types of explanations come up soon or later: either farmers are "doubtful" about the innovation and then "cultural" or "psychosociological" reasons are invoked (*infra*), or it is the extension system which is being questioned, e.g. its authoritarian nature. But the validity of the technical message, i.e. the pertinence of the innovation, is not questioned when the problem often resides at that level.

One of the tasks of agricultural technology – a science which Sigaut considers as *a branch of comparative agriculture* – would be "to shed light on this whole unclear area made up of internal knowledge, i.e. knowledge actually elaborated and transmitted by farmers, outside any voluntary intervention by advisory bodies"

[4]National Museum of Natural History.

[5]On the issue agricultural innovation, see the work edited by Chauveau et al. (1999).

(1976b). More than 30 years after Sigaut's invitation, accepting this knowledge has progressed a lot fortunately, and many works have been carried out by geographers, anthropologists or ethnologists as often as actual agronomists.[6] A certain rehabilitation of farming practices has led to replacing innovation problematics at the centre of works and research conducted on rural development (Chauveau 1999). All recent approaches to innovation management in agrarian dynamics, are giving way to the historical depth of the processes on the one hand, and to their qualitative analysis on the other. Innovation is no longer considered as being linked to an advisory package "distributed" by extension agents, but as a complex process (idem). Furthermore, the recent report of the International Assessment of Agricultural Science Knowledge and Technology for Development (IAASTD), an international group of experts on agricultural sciences and technologies at the service of development, offered major expertise on the issue, illustrating the consensus that was recognised from then on within the international scientific community, on the potential of "local knowledge" and "traditional techniques" as regards sustainable development (2009).

In French rural areas, it is often striking to note the proximity between certain production systems "condemned to disappear" because they are insufficiently equipped, too confined, "little technified" and described as "little innovative", and the technical specifications of organic farming or certain labels, all signs of quality put forward as *innovative* and bearing a more sustainable development. Paradoxically, the official recognition of these quality signs often remains inaccessible to farmers implementing systems that are yet very close. So-called "precision" agriculture offers another example where innovation might not be where we think. With the new equipment, one is in a position to apply adequate quantities of fertilisers and pesticides with precision to heterogeneous plots, an innovation which is presented as major as far as sustainable development is concerned.[7] But plots treated as such measure several dozens of hectares and, after being consolidated, result from several dozens or even hundreds of plots being grouped together, each one being in the past the subject of a specific treatment, with the adaptation of quantities of manure and different treatments with micro-local characteristics. And so precision is *back*, but on a scale of vast and standardised plots; and innovation is not agronomic in the strict sense but, rather, initiated by farm equipment firms. Waiting for this new equipment to one day become more affordable, today its utilisation is limited to a small number of farmers because of its cost, i.e. those very farmers who pushed the "productivist" contemporary revolution process farthest, and who contributed the most to the standardisation of the landscapes and to the elimination, on the face of it, of intraparcel heterogeneousness.

[6]Too many of these works exist for them to be quoted here. An example of it is given by the compendium of contributions presented by Georges Dupré (1991). See also, for example, the works of Pierre Milleville (2007) and those of Paul Richards (1985).

[7]This led a cereal farmer from the Aube *Département* (with 400 ha) to be awarded a "sustainable development" medal during the Paris International Agricultural Show in 1999.

As far as "sustainable development" is concerned, because "project leaders" often communicate better and more than others, do they not make certain practices out to be innovative, practices that, yet, were already old but that were despised or denied for so long that their simple recognition, as such, would be frightening? The debate established in France around taking into account "what already exists" in land farming contracts, i.e. the recognition, by the public authorities, of *old* practices (already in place) but deemed favourable and compatible with the spirit of land farming contracts, is a good illustration of this questioning. The outcome of the debate was that one had to "do more", i.e. accept this conception of innovation as necessarily coming from *elsewhere*.

Today's debates on sustainable development in general, and on agro-ecology and "ecological" intensification in particular, actually shed new light on the issue of innovation (Griffon 2006). Issues developed around biodiversity and local knowledge in this regard are considerable, as shown by the endeavours of certain industrial groups to appropriate this knowledge, and by international discussions on the matter. The race for innovation is transforming into a race for ancestral knowledge. But the paradox remains: while an increasing number of agricultural development researchers and practitioners would like, from now on and with raison, to see the functions brought by the actual ecosystems in productive processes mobilised as far as possible, agronomists are the ones who are expected to act, even though the desired innovations are already very largely implemented by some rural societies, these being yet largely ignored. This is the case for example of the biological complexity of cropping systems established in many agrarian systems, a complexity that sometimes compensates surprisingly well the complete lack of industrial and fossil energy inputs.[8]

Furthermore, the spheres of government often contribute to choosing the direction to be privileged as regards innovation. The case of the maize-soya "model" in animal feed has already been mentioned. That of the eradication of slash-and-burn agriculture in South-East Asia was deciphered by Olivier Ducourtieux (2010). The case of the extension of hybrid maize in some regions of Africa or that, more recent, of NERICA rice could also have been influenced by the authorities. We are not addressing the issue of genetically modified organisms here, as it has been widen taken over by the public debate. Such questionings would call for more developments, especially for the emergence of a new discipline, political agronomy, in the image of the works recently undertaken by Sumberg and Thompson (2010) in *Contested Agronomy*.

[8]Patenting the use of termites' nest powder as bio-fertiliser for market gardening, as announced by the IRD in its periodical Sciences au Sud (n°40, July–August 2007), even though most farmers of the intertropical zone already master, beyond what we could imagine, the biological functions brought by these animal communities, also illustrates this phenomenon, in addition to the issue raised by copyrighting this type of innovation.

Chapter 8
Making History in Comparative Agriculture

8.1 A Problem of Sources

France is undoubtedly one of the countries in the world where rural history has been studied the most. Beyond the great names of that history, Marc Bloch (1931), Michel Augé-Laribé (1955), Emmanuel Le Roy Ladurie (1969), or George Duby and Armand Wallon (1975) to only mention these names,[1] there are indeed many works carried out by historians, or geographers, on the landscapes and agrarian structures of Western Europe in the nineteenth century and the first half of the twentieth century, landscapes and structures where old history was perceived as very meaningful. On other continents, we could mention for example the works of François Chevalier, Enrique Florescano, Jean Piel or John V. Murra on Mexico and Andean America, or those of Jean Pierre Chrétien, Catherine Coquery-Vidrovitch and Elikia M'Bokolo on Sub-Saharan Africa, all indispensable inputs in initiating the historical reconstruction of former agrarian systems.

However, even if historians' input is significant, especially in a country like France as far as old periods are concerned, the working methods developed by history on the one hand, and comparative agriculture on the other end up diverging, particularly as regards sources used for more recent periods. In countries where historiography is abundant for old periods, there is very often a large gap between this often very rich corpus of knowledge, and that, much more incomplete, dedicated to the contemporary period.[2]

---

[1]Indeed, we would also need to mention Jean-René Trochet (1993), Jean Marc Moriceau (1999), Jean Luc Mayaud (1999), Philippe Blanchemanche (1990), and many others... The agrarian history manual *La terre and les paysans aux XVII and XVIIIè siècle*, (1999), edited by Jean Marc Moriceau, testifies to the wealth and diversity of these inputs.

[2]In France for example, it is the geographers rather who have "taken care" of the recent history of agriculture, the evolution of structures and the rural exodus issue in particular. Jacqueline Bonnamour offers many examples of it (1993).

H. Cochet, *Comparative Agriculture*, DOI 10.1007/978-94-017-9828-0_8

Where statistical sources become increasingly abundant and detailed for these periods, studies on the recent history of agriculture often make of statistical *series* their basic material, with the more *localised* sources (archives of different nature, maps, surveys) being considered of secondary importance. This change in source and scale of analysis is not without problems. Concerning the use of statistics and without elaborating on the issue of their lack of reliability in many countries, the fact that statisticians simply choose statistical *units* and *categories* adopted for the analysis, makes their usage tricky in comparative agriculture. Concerning statistical units, for example, the administrative division usually adopted (*commune*, *canton* and *département* in the French case) rarely corresponds to landscape units or regions that would be pertinent from the point of view of dominant forms of exploitation of the agro-system, from the point of view of *agrarian systems* (*supra*).[3] Concerning these adopted statistical categories, many difficulties arise, for example when reading the RICA, when the "techno-economic orientations" (the OTEX), a simple category for statistical use, are considered as *production systems*, yet a very different notion (*supra*). The statistical categories adopted sometimes also reflect a certain *comprehension* of history, and therefore of social relations which must be deciphered. In Mexico for example, the revolutionary and agrarian ideology dominating historiography after the Revolution of 1910, made of the opposed categories of *haciendas* and *peones* the only figures of rural areas, to the extent that only these categories were adopted in the statistics of the time. Tenant farmers, although in great numbers, disappeared altogether from official documents; the agrarian reform was not intended for them. Likewise, because *latifundia* were declared illegal – on paper – in 1934, they were from then on registered into the statistical category of "small properties", as if to confirm further the myth of their disappearance. Therefore, statistical categories are sometimes misleading and not very adapted to our potential use of them. Another example of the flimsiness of statistical categories is found with the *separate* census of areas sowed with different crops (maize, beans, marrows etc.) in countries where intercropping prevails widely.

Finally, it is rare to find in other regions of the world a country with a historiography as rich as that of France, which means that working on comparative agriculture is made all the more difficult. In the case of Sub-Saharan Africa, and because the very historicity of these societies was denied for so long, the history written by historians is still very much incomplete. Of course, contemporary historians who worked on the pre-colonial period brought *oral sources* to the fore, thereby compensating for the lack of written sources dating from that period. A significant survey work could then sometimes be carried out, and a pre-colonial history could be "re-established" (Chrétien 1993). Breaking away entirely from

[3]Far from relying on the reasoned reading of rural landscapes, the new administrative divisions carried out in France after the Revolution, deprived statistical data of part of their meaning (data on the natural environment and farming activities for example). Later on, a remarkable effort was made in the country to divide the territory into "agricultural districts". The pertinence of this, which had been questioned for a long time in favour of "production basins" in particular, has been revived today around environmental problematics and "country" approaches.

colonial literature and ideology, this critical approach, implemented by African and European historians, gave their voice back to Africans and breathed new life into pre-colonial history. It contributed to questioning "the failure to act" of agrarian societies, and to endow them with a *historicity* at last recognised, an essential prerequisite for understanding contemporary dynamics, particularly from the agricultural point of view. This approach also highlighted the innovation and progress capacities of the concerned populations, as well as true endogenous (political, demographic and agricultural) development dynamics. This historical knowledge which was finally extended to the pre-colonial period, also led to envisage development, not as a simple process beginning with Africa coming into contact with the "civilised" world, but as the outcome of European and international dynamics on the one hand, and changes internal to African societies on the other (Couty 1981).

However, this "new" history frequently falls back into the failings of the past when historians, taking an interest in the colonial period, stop resorting to surveys as soon as the abundance of written sources (of colonial origin in particular) push rural testimonies into the background. Although they very often only reflect the viewpoint of the coloniser, even if it were appropriate to nuance this affirmation, written sources remain more reliable than oral sources which, besides, are much more difficult to record.

Other disciplines also contribute to deciphering changes in societies for which few written documents are available. This concerns *Environmental History* in particular as mentioned earlier (Sect. 3.5), with works shedding light on the complexity and dynamics of relations between farmers and their ecosystems (Tiffen et al. 1994; Fairhead and Leach 1996; McCann 1995, 2005). For older periods, inputs from archaeology, palynology and linguistics sometimes prove to be decisive.

Despite history and other previously mentioned disciplines contributing significantly to knowledge on old transformations in agriculture and rural areas, the field work and detailed interviews with the main actors of this history, elderly farmers in particular, remain crucial in exploring the more recent periods still accessible in the collective memory and, as seen previously, less frequently dealt with by historians.

Furthermore, there is the issue of the research subject and concepts used as analytical basis (*supra*). Here again, what we need to search for and find in comparative agriculture, most often differs and deviates from the objectives pursued by historians. Rather than being limited to using data that was neither conceived nor collected to answer the questions posed by comparative agriculture, and to tackle its objective, it is often better to build up a proper and sufficiently consistent collection.

8.2 Historical Surveys in Comparative Agriculture

"Writing" history should not be limited to reconstructing a series of events, as determining as these are. Through this approach, one should be able to understand how people experienced these events, i.e. how their practices were or not modified by them. One should be able to relate the different periods marking the evolution of agro-pastoral activities, to characterise them and to explain their underlying causes. In order to go beyond all knowledge produced by ruralist historians, and to follow more specifically our object of study, i.e. the old and contemporary transformations of agriculture, a historical method peculiar to comparative agriculture needs to be developed. Forged on the practice and know-how developed in extremely contrasted historical and geographical contexts, this method is still relying on field work: landscape analysis and oral surveys.

In comparative agriculture, the idea is to try, through detailed interviews with farmers (the older ones in particular), to identify concrete facts relating to agricultural and livestock farming activities, and to think about potential links between these different elements. To arrive at concrete elements that can be verified by repetition and cross-checking, starting from the concrete and relying, with our interlocutor, on a material basis is another matter. The landscape emerges once more as the point of departure of the analysis. Marc Bloch himself was relying on the landscape when he wrote: "Interpreting rare documents leading us to understand the obscure genesis of rural landscape, formulating problems accurately, and even having an idea of it, a first condition had to be fulfilled: observing and analysing today's landscape. [. . .] here as in anywhere else, change is what historians want to grasp" (Bloch 1949).

It is on the basis of a functional analysis of the landscape, and on the location of its different parties and their main constituents (an analysis that must be carried out beforehand), that it becomes possible to start a discussion with an elderly person on the history of the landscape. With a careful observation and "reading" of this landscape, it is possible to gather many visual and factual elements on practices, and to formulate a number of interpretive hypotheses on the "functioning" of the landscape and the systems forging it, as well as on the most recent modifications with traces still perceptible. Indeed, the agrarian landscape can be likened to a large book opened on the last page, where one would be trying to flip through the previous pages. The main difficulty is that these pages are most often stuck to one another, and more or less opaque or transparent, so much so that the whole appears like a set of veils covering one another fully or partially. "Only the last film is intact; to recreate the broken features of the others, it was necessary to first unwind the roll in the opposite direction of the filming" (Bloch, op. cit.). In this regard, the image of an incomplete palimpsest has sometimes been brought up by geographers. Sautter wrote: "Landscapes are never the pure expression of a set of forces in action. They are always mixed with a portion of physical and human legacy. This is actually what makes it possible, from surviving traces or elements, to recreate the landscapes of the past. The mixed status, half-way between past and present, prohibits functional simplifications" (Sautter 1985).

How can one make out the most recent elements, those that existed immediately before, and evidence of elements even older but almost completely erased today? In order to succeed in recreating the presence or absence of one element or another at each step of the history of that landscape, one needs to permanently keep in mind to locate each element with precision, first in time (what era are we talking about?), and then in space, i.e. in an ecosystem (or an agro-eco-system).

It is by being confronted with this landscape, and by dissecting it one element at a time with one's interlocutor, that it becomes possible to place each gathered information back into space, all the more since visualising the elements used in supporting the conversation (an old stump, stones in a field, a ravine resulting from erosion, a trimmed tree, a pit or an embankment, among others) stimulates and revives people's memory. What is that tree? Who planted it? Did you see it being planted? Were there others like it? And what was being done with them?[4]

In order to place all these elements back in time, one needs to work first in relation to the actual reference points of the interviewee. Indeed, mnemonic work is organised according to a personal framework, punctuated with the events and/or periods that have marked the most the interlocutor's life history. This framework must be recorded beforehand from the interlocutor, so as to be able to refer to it throughout the ensuing conversation, and as such replace each recalled element into a chronological scale. Where various interviewees often share only a few chronological reference points, this means that a long cross-checking and summarising work must be conducted on the basis of another periodisation, this time peculiar to the researcher, and in conformity with the previously established corpus of hypotheses.

Because this mnemonic work calls on the actions of the elderly at different periods of their lives and because they try to piece together, in front of today's landscape, the space of people's daily lives, these interviews must be carried out individually for accuracy. Indeed, there is nothing more difficult than to locate accurately, in time and space, a mishmash of information from a collective discussion gathering several people who did not have the same experience, with each person making reference to different spaces and unmatching periodisations. This would lead to generalities which, at best, would represent faithful descriptions of "average" practices, but which would be difficult to locate accurately in time and space, and as a result would be of little use.

Furthermore, this approach to history can also be clearly distinguished from that implemented by historians specialising in the African continent. Indeed, since oral sources began to also hold the attention of Africanist historians, there has been a large consensus on recognising the importance of gathering "oral traditions" from elderly people, and on retranscribing these. However, the historicity of these traditions has been the subject of debates. While they are often formalised and intentionally transmitted, oral traditions are always constructeddiscourses reflecting

[4]Resorting to old air photographs, when they exist, can facilitate this reconstruction work considerably. They can then be used during conversation and give rise to a fascinating joint reading.

the "public text" readily dished out to visitors, and expressing false unanimity,[5] even though contradictions internal to the group or expressions of disputes would also represent rich sources of learning.

Collecting people's agricultural and pastoral practices through interviews with elderly people is not without risks. To take advantage of oral sources, one still needs to reinterpret what is being said in terms of what has been experienced and according to the circumstances of this recollection, and especially in terms of the social status occupied by the interlocutor[6] at the time (or today still).

In countries where one or several periods of war and/or authoritarian political regimes have left bad memories, visitors initiating a conversation on these periods with members of the public or people in positions of power, have a hard time collecting information other than value judgments, expressing condemnation of the "old regime" ideology of the period concerned. It then becomes particularly difficult to go beyond collecting this discourse and gathering precisely the elements required to reconstruct a mode of exploitation of the environment, a production system or a relationship with the land. The rigorous reconstruction of agrarian transformations in the countries of the Soviet Union, immediately before and after the transition in particular, is a very long process. Yet it is crucial in apprehending the contemporary agricultural dynamics in this part of the world.

Another example is also found with the recent agrarian history of Ethiopia. While it is true that the Derg[7] left very bad memories for many people, it seems essential to achieve a finer periodisation of the "Derg era" rather than, during the historical interviews conducted with people who lived during that period, stick only to a before/during/after-type of periodisation, which is far too basic and likely to result in simplifications. Indeed, it seems necessary to try to assess *separately*, on the one hand, the application methods, on the ground, of the first agrarian reform measures taken in 1975 and their consequences until 1979 (abolition of the social relations of the former regime, peasant agrarian reform) and, on the other, the collectivisation (from 1979) and villagisation (from 1985) policies and their differentiated effects on peasant farming. In many regions of Southern Ethiopia, it was the second and third periods in particular (collectivisation and villagisation), which left a negative mark on the memory of people, and which sometimes left a long-lasting mark on the landscape (abandoned old dwellings, fallow lands, ghost villages with locked houses, etc.). But it was during the first part of the revolutionary period (1975–1979) that, with the abolition of the social relations of the former regime, the foundations of the current land structure had in fact been laid by making the triumph of small individual farm tenure complete, despite succeeding attempts at collectivisation. Let

[5]The "public text" is opposed to the "hidden text" which is never revealed publically by dominated groups, in that it contests the established order and is reserved for the intimacy of the limited group (Scott 1990).

[6]Dupré showed for example the difficulties and dangers of piecing together the "original vegetation" through interviews with farmers (Dupré 1991).

[7]Authoritarian regime of Marxist-Leninist inspiration headed by Mengistu from 1974 to 1991.

us recall that all the "post-Derg liberalisation" did was to reinforce the benefits of the first agrarian "peasant" reform of 1975, without modifying its basic principles (fundamental State ownership and inalienable farm tenure) (Cochet 2008b).

Whether it is about carrying out the agro-economic "diagnostic" of a given region, or about thinking in terms of development projects and policies, comparative agriculture practitioners are still questioning one thing: how far does one need to go back to explain the present? One needs to know where to stop when on such a counter journey. Irrespective of the cognitive importance of the study of very old periods, one needs to explain the present, and not to "do history for the sake of doing history". In France, all it takes in general is to go back to the beginnings of contemporary agricultural revolution, i.e. soon after the Second World War, to perceive with sufficient precision the paths taken by the accumulation and modes of differentiation of the production systems, since the explosion of the mixed crop-livestock farming systems prevailing at the time. And yet, one must still be able sometimes to detect, in the landscape, and subsequently confirm through interviews with elderly people and appropriate readings, traces of older changes in resource allocations and productive combinations, changes that contribute to shedding light on the present: converting ploughed lands into permanent grasslands in Normandy and Charolais, giving up ploughing on the wet lands of the humid Champagne area or the Liassic lands of Burgundy, following motorisation.

In all the countries that went through important agrarian reforms during the twentieth century (Latin America, the former Eastern Bloc, China, Vietnam, Ethiopia, etc.), it would be unrealistic to try to understand anything about contemporary agricultural dynamics without reconstructing the agricultural dynamics inherited from these agrarian reforms. In order to perceive their origin and make a real assessment of these agrarian reform policies, one needs to know with a minimum of precision what was the situation like *before* the disruptions, which sometimes brings us fairly far back. The same requirement applies to Sub-Saharan Africa where too many "developers", disconcerted by the complexity of reality, ended up calling now and again on indigenous "mentality", the "cultural factor" or simply their laziness, to clear their names, at little cost, of their ignorance of the past.

However, as fascinating as they are, historical interviews do not lead to collecting more than a series of data, a set of elements located in time and space, which could actually even be their objective. How can one make historically significant *facts* of these elements? And what about the sequence of these facts, what can be said about their functional and systemic relations? How can one identify and add the missing pieces of the puzzle to then reconstruct an *agrarian system*? As such, we assess that, without hypotheses, there can be no pertinent survey no more than there can be a comprehensive reading of the landscape, and that without clear concepts, there can be no hypotheses either (*supra*).

By escaping the impossible and unrealistic exhaustiveness of the survey, we then carry out a reconstruction exercise, and in the process, a reconstruction of cropping systems and livestock systems practiced by some at different moments of history and on different sections of the ecosystem, a reconstruction of production systems, and finally a significant agrarian system outline. This intellectual construction, relying on a set of concrete elements and located in time and space, but at the same time inspired by a set of precise concepts, is after all nothing more than an exercise in *modelling*. For the result of this construction is looked on as a model, a mental construction, a view of the mind some will say, admittedly a simplifying outline, but one that leads to progressing in understanding reality, itself always more complex, diversified and contradictory than the model supposed to represent it.

Chapter 9
How to Build Farm Typologies?

9.1 Brief Survey of Typological Methods[1]

In the domain of agricultural development, many research works in the 1970s and 1980s, focused on the systemic modelling of farm operations and typological methods. While many researchers became increasingly aware of the complexity of the rural world and the necessity to understand its diversity for the sake of development, they set themselves the target of explaining why all farmers within a region did not react the same way to technical advice, innovation etc. The target consisted in building typologies highlighting differences in farming means and operation, and was concerned with classifying farms into a limited number of relatively homogeneous and contrasted categories: a sufficient number so that differences were not too wide, but not too important so that typologies remained usable. The idea was to understand the dynamics and functioning of the farms in each category, as well as to compare and explain their differences.

The search for "differentiation criteria" then represented the most widely used entry point to building typologies, i.e. a means to apprehend and classify farm diversity. But how were these criteria to be organised into a hierarchy? While by virtue of a first criterion deemed discriminating, e.g. the size of the farm, three classes are created, the intervention of a second criterion deemed important, e.g. the use of a salaried workforce, rapidly leads to six farm types, unless some of the boxes of this double entry table (what do we do at the third criterion, or the fourth?) remain empty. The necessarily very low number of criteria adopted and their prioritisation are indeed very likely to remain widely arbitrary . . . or dependent on the point of view of each one. As such, one could classify the farms of a region in many ways, depending on the criteria being adopted and the ends being pursued.

[1]For a brief survey of typological methods, see Cochet and Devienne (2006). For the case of Sub-Saharan Africa, see Jamin et al. (2007).

H. Cochet, *Comparative Agriculture*, DOI 10.1007/978-94-017-9828-0_9

Also, there would be as many possible typologies as objectives ascribed to each one, as proposed by Jouve (1986). With the incurred risk that none of these typologies actually lead to identifying and classifying *production systems*, and even less to identifying their paths, measuring their performances and future perspectives, and understanding the dynamics of an agricultural region, how can a production system typology be elaborated in such a way as to be generally applicable, and be free of its subsequent uses, a progressive typology with cognitive function?

Another approach developed at the end of the 1970s, placed farmers' "objectives" at the centre of the typological approach. The production system being considered as depending on the farmer's long term project or still "finalised" through the farmer's objectives, one then thought that it was sufficient to identify these "objectives" in order to find the systems and as such build a farm typology.[2] The farmer's objective and therefore his age (the former often being conditioned by the latter in the mind of many authors), emerged as the entry point in the diversity of typologies, and as a new method for their elaboration.

However, this entry via the farmer's "objectives" in turn creates a problem. Would all the farmers who wish to establish their son – this is typically a "long term project" or a clearly expressed "objective" – implement the same production system in order to reach this objective? Likewise, the desire to reduce the hardness of the work or to be able to take a holiday can of course influence the farmer's decisions, or even lead certain investment decisions in a specific direction. Yet, the farmer could not determine on his own, even partially, the combination of the production factors and activities peculiar to the farm. If this desire to take a holiday surfaces at a given moment, to the point where it becomes a *declared* objective of the farmer, isn't it just as much because, today, the characteristics of his farm and its evolution make the expression of this desire possible? Are we not likely to invert causes and consequences too hastily? Would not the neighbours, who are still forced to milk 25 cows shackled in old buildings, using a milker pail, also want to take a holiday, even if the idea of formulating such a wish – the objective – never came to mind? Likewise, what to think about the "extension" objective as the key for identifying a production system or, worse still, about the "disappearance" objective to designate declining farms?

9.2 Prior Identification of Production Systems

Rather than searching in vain for the "right" differentiation criteria (necessarily in very limited numbers, with their selection being rarely devoid of arbitrariness, i.e. so many points of view, so many typologies etc.), or leaving it to the only objectives

[2]This approach saw the light at the end of the 1970s with J. Brossier's and M. Petit's proposal: "On a Typology of Farms Founded on Farmers' Projects and Situations" (1977). See also Capillon and Manichon (1979) and Jouve (1986).

declared by farmers, a cognitive farm typology must first of all rely on a *prior* identification of production systems. The idea consists in identifying production systems before initiating a detailed study of how they work.

By identifying production systems beforehand, one avoids falling into the trap of typologies deduced from the classification of (farm) individuals by type, according to certain "differentiation criteria" (structure variables for example) determined beforehand.[3] Rather, this identification leads to *building* ideal types and endeavouring to seek for each one of them a maximum of logical coherence with an explanatory purpose (Perrot and Landais 1993). Each type is then characterised by a set of attributes peculiar to it and leads to understanding the internal logic of its existence, its specific operation as well as its trajectory. Consequently, there is no reason to seek to build a unique set of variables for collection, carefully organised in a pre-established standard questionnaire.

With this method, one must be able to choose the farms to be studied in detail according to the production systems identified, i.e. to carry out a *judgment* sampling, and in parallel, to be able to ask pertinent questions, so as to characterise the production systems accurately and understand the reasons behind farmers' choices (Cochet and Devienne 2006).

Resorting successively to a reading of the landscape (*supra*), then to an analysis of the historical transformations of agriculture in the region, leads to formulating preliminary hypotheses on the elements which in turn lead to locating and explaining farm diversity (Dufumier and Bergeret 2002). Current production systems, their differentiation as well as their diversity are the product of a historical dynamics – or a trajectory – which it is absolutely necessary to reconstruct with care. The current state of this differentiation is the *product* of that history.

Would the reasoned decision to tackle identification through history lead to historical determinism? We will object with reason that farmers' decisions are very far from being influenced only by their career path, and that the history of the farm cannot on its own explain its current state or the farmer's future choices. But this is coming back to the specific case of one farmer or another, to the latter's decision to stop milking so as "to be able to take a holiday", or to his neighbour's decision to ensure conformity "so as to establish his son", to the detriment of a regional approach to production systems, each system being then understood as a *model* representing the operation of a *set* of farms located in comparable (agronomic, economic and social) production conditions (*supra*).

In this perspective, and for a given farm type characterised and modelled by *a* production system, individual investment – and therefore evolution – choices, are then *necessarily* part of a restricted field of possibilities open at a given time of its history to this farm type. The conceivable "trajectories" for the farms of a region are therefore in limited numbers; they illustrate the result of the differentiation mechanisms at play for each historical step of transformation of agriculture. That is why it seems to us that finding these differentiation mechanisms

[3] Method used for so-called automatic classifications.

and these trajectories, is the best way to carry out the efficient and cognitive identification of production systems existing in a region, far beyond a criterion deemed discriminating or a declared "objective".

Resorting to history and to the differentiation mechanisms of production systems, also has the merit of identifying "fast disappearing" production systems, and even those that have already disappeared, with their traces in the landscape busy fading (in this case reading the landscape also turns out to be valuable). All are systems too often barely identified and rejected here and there in the encompassing "traditional" or "little technified" category. Yet finding these systems and their detailed characterisation, identifying and understanding the phases, causes and mechanisms of their disappearance and its consequences, are extremely useful in detecting differentiation mechanisms, and in understanding how the other systems could transform and be what they are today.

As such, a careful reading of the landscape and a detailed reconstruction of the history and transformations of regional farming via historical interviews with elderly people, particularly with farmers who are retired or close to retirement, represent in our opinion the two pillars of the true identification of production systems, an identification which precedes their detailed characterisation and which is on the scale of their economic performances (Cochet and Devienne 2006). We will then proceed with a judgment sampling of production units to be studied in detail, so as to apprehend the diversity of situations and favour the comparison of the technico-economic processes and results.

9.3 Taking into Account the Conditions of Access to Land and Resources

The conditions in which farmers manage to gather the land resources they need, as well as the required irrigation water, production means and workforce, influence directly, as was recalled concerning the production system concept (*supra*, Sect. 3.3), the established productive combination. That is why taking into account these conditions of access to production factors, is often essential to the pertinent identification of production systems and their subsequent characterisation. Furthermore, they partly influence the rules for sharing out value added and therefore weigh a lot on the elaboration of the farmer's income (*infra*).

There are even cases where differentiation internal to rural societies, focuses on the search for a production factor, whether from the point of view of the social organisation or that of the productive processes. This is the case, for example, of certain agrarian systems characterised by access to land regulated by kinship, a very low level of capital and manual tools, and where everything (almost) relies on the manual labour of farmers and their families. Households likely to "attract" a workforce from outside the domestic group, particularly during the peak periods of farmwork planning, can establish production systems that differ in many ways from households that, on the contrary, are forced to let others have part of the group's workforce. Consequently, it is indeed access to this production factor, i.e. the workforce, which must guide the typological approach.[4]

Very often, the productive choices being implemented ensue from the capital/labour internal to the production unit. That is why, in some situations, it is essential during the elaboration of typologies, to identify the different "categories of farmers" involved in the agricultural development of a region, before examining the production systems farmers are likely to implement (Dufumier and Bergeret 2002). Depending on whether farms will for example be of the "minifundia", "commercial family", "employer's" or "capitalist" type, the odds are that they will not benefit from the same conditions in selecting productions as well as cropping and livestock farming systems to be developed, and that consequently this will result in the establishment of very different production systems.

9.4 Explaining Diversity

In order to explain farmers' choices and thus give meaning to the diversity observed within an agricultural region, approaches based on the construction of a typology leaving an important place to farmers' "objectives", as an explanatory factor, focused once more on the operator as subject, thereby questioning "materialism" or "economic determinism". Placing the operator at the centre of the analysis on diversity, and making of him a subject with "free" choice, undoubtedly leads to apprehending part of the diversity on a micro-regional scale, although, as seen already, one's choices appear greatly determined by one's legacy and all the things that are "possible" at that moment.

But the choice which is still possible at the level of the operator is almost no longer perceptible, nor does it take on an explanatory nature, as soon as we change the scale of analysis on diversity. As such, the approach to farmers' diversity cannot content itself with this scale of analyse, that of the "subject individual". It seems that we need to tackle two much larger, temporal and spatial dimensions in trying to understand diversity, that of long term history on the one hand, i.e. that of intergenerational dynamics that have built the differential evolution of trajectories and systems, and that of the geographical differentiation of agrarian systems on the other hand. As soon as farming diversity is not reduced to the variety of "opinions" on agriculture, as soon as it is not limited to a diversity resulting from the free choice of operators and to representations that can explain or justify these choices, and finally as soon as we try to understand the diversity of ongoing processes and their evolution, we must strive to build meaningful cognitive typologies. In this sense, the typological method implemented in comparative agriculture brings differentiated trajectories to light, explains mechanisms presiding over this differ-

[4]Many examples could be mentioned in Sub-Saharan Africa.

entiation, explains in detail the existing relations between the different categories of farms (flow of workforce, biomass, capitals etc.), highlights the differentiated impact of the production systems on the exploited ecosystems, shows how policies and projects did not have the same effects on the different production systems, and finally anticipates the future dynamics peculiar to each category and their consequences as regards production, employment and environment, as well as the possible and differentiated impact of changes in agricultural policies.

Comparing production systems within an agrarian system, also leads to understanding the overall coherence of the agrarian system, the regulation mechanisms founding this coherence as well as its internal contradictions. The validity of the comparatist approach on that scale is not an issue as such, its explanatory virtues having been widely shown. Choosing to work on the scale of the agricultural district is also validated, because this comparison must imperatively embrace all present forms of agriculture so as to detect the overall logic and the dynamics at play, and not just one category of farmers or another that would be more specifically the target of an agricultural policy or extension programme.

Chapter 10
Economy of Agricultural Production Process

10.1 Linking Economic Approach and Technical Processes

Comparative agriculture was born in a *Grande École* of agronomy. The existence of these very French higher education institutes and the fact that their identity was for a long time formed away from universities, probably explain in part the existence in France of a somewhat original population, that of agricultural economists, i.e. scientists and professionals who, coming from these *grandes écoles* of agronomy, are equipped with a basic "agronomic culture" and are later on trained in economics. This population is fairly clearly distinguished from rural economists who are equipped with a basic training in economics, usually come from universities, and are much later on made aware of a specific activity sector, agriculture in this case. Outside of France, it is often this second category of economists which dominates the field of agricultural or rural economics. Yet, the confrontation of these two populations in the field proves to be more difficult without appearing to, and although their disciplines are closely related, they can still decide not to listen to each other... While the first are often considered by the "real" economists as "second rate" economists with a poor economic culture, agricultural economists criticize their colleagues for not being able to reason with *technical issues* and for their lack of relationship with the field... while the ones delight in elaborating equations that have sometimes nothing to do with reality, the others limit themselves to low-key empiricism... And when both try their hands at transgressing these invisible boundaries, the result is not approved unanimously by either side; scientific objects, used sources and working methods differ too often.

While there is no doubt that comparative agriculture is situated rather in the loose conglomeration of agroeconomics, what more can be said about the relations, proximities or differences between Comparative agriculture and "Economics"? And which economics are we really talking about?

H. Cochet, *Comparative Agriculture*, DOI 10.1007/978-94-017-9828-0_10

Irrespective of the historical period and the region under study in comparative agriculture, *production processes* and their evolution are at the centre of the analysis. Once these processes have been identified, thanks in particular to the production system concept (*supra*), and once the technical *operation* of these systems have been carefully analysed, one of the tasks of the agricultural economist will be to evaluate the economic results and performances of this operation.

In comparative agriculture, the economic approach at the level of the production system, must necessarily strive to link the economic results of the system with the constraints of its technical *operation*. On the one hand, the economic results of each production system depend on its technical operation; on the other, the economic calculation is essential to help shed light on the operation, to understand why within a region farmers practice different production systems, and to offer hypotheses on the perspectives of farm evolution. This interface between "technical" and "economic" dominates our approach to the production system concept, just as it takes us slightly away from the usual evaluation criteria of general accounting, as will be seen later on.

10.2 Economy of Agricultural Production Applied at the Production System Level

It is at the level of analysis of the farm, and therefore of the production system that the measure of economic efficiency of production processes is the most interesting, and also that comparative agriculture took over a few basic economic concepts. In order to measure the economic performance of farms, to assess the efficiency of the farmers' work at that level, and to compare these results from one farming group to another and from one region to another, three economic magnitudes are particularly interesting to study: *value added* (VA) which expresses the creation of wealth resulting from the system's operation, *productivity* which measures the efficiency of the production factors, labour in particular, and *farm income* understood as resulting from the distribution process of value added.

Value Added and Productivity

Let us recall that net value added measures the *creation of wealth* of the production system. It is equal to the difference between the value produced (the gross product) and the value of the goods and services consumed in whole or in part during the production process. In order to carry out a calculation that accounts faithfully for the concrete operation of the production system, gross product (the value of final productions including home consumption, measured at market prices) and intermediate consumptions can be evaluated directly per crop or per unit from the yields, from the prices of the different products and from the crop management sequences or herd management patterns, therefore from the technical operation of the production system. As to fixed asset depreciation (or amortization on replacement value), it is evaluated on the basis of its real utilisation period, a period which is here considered as a characteristic of the production system (Cochet and Devienne 2006).

Moreover, the relevance of the notion of value added is to lead to a comparison, between production units, of the economic results obtained, irrespective of the methods used to distribute value added between the operators who contributed to its creation. Whether the production unit is family-based (with the result of its operation in the end reflecting a farm income) or a capitalist-type company (where profitability is privileged), whether value added remains mostly in the hands of the producer as the owner of the land and means of production, and works with the family labour; whether it is distributed between the farmer, the landowner, the banks and the salaried workers, or on the contrary concentrated into the hands of the capital provider (*infra*), value added remains the universal criterion that actually initiates a comparison of economic performances of the different forms of agriculture found today around the world.

While the use of the concept of value added does not create a problem, beyond of course the precautions to be taken at the level of data collection via interviews with farmers, the use of the term "productivity" gave rise to various interpretations in agriculture.

In economics, this term refers to the relationship between value added (the difference between the value of goods produced and that of goods consumed during the production cycle) and the quantity of production factors used to produce them, capital and labour in particular. We speak of "capital productivity" to refer to the relation of value added with the quantity of fixed assets, and of "labour productivity" to refer to the relation of value added with the quantity of labour used (measured in hours or days of work, or still in number of labourers). "Global productivity" is the relation between value added and all the production factors used, i.e. capital and labour.

As soon as agronomists and agricultural economists began to use this term, they made the range of the use of the term "land productivity" richer, in order to take into account the production factor specific to the agricultural domain. Referring to value added produced per hectare, "land productivity" is used widely today by agronomists, as much, if not more than "labour productivity".

In order to measure the economic performances of farms, to assess their efficiency and compare the results from one group of farms to another and from one region to another, both *labour productivity* and *land productivity* are of course essential. While land productivity (i.e. annual value added in relation to the total area of the production unit) is an expression of the result of the intensification of the productive process, labour productivity (annual value added in relation to the quantity of work carried out) measures the efficiency of the labour incorporated into the productive process. However, there are several ways of measuring this last magnitude, depending on whether the quantity of labour necessary is expressed in hours or days of work on the one hand, or in man-work unit (MWU) on the other. In the other sectors of the economy, a year of work can often be considered as a simple multiple of the hours or days of work, which makes this distinction pointless. However, in the agricultural sector, where work is most often seasonal, the two approaches of labour productivity offer different and complementary results. Value added measured for an agricultural labourer and per year, measures the economic efficiency of a labourer in a given production system, and as such is an expression of *global labour productivity*. On the other hand, value added related to a day (or an hour) of work, i.e. *daily (or hourly) productivity* of work, introduces (1) the economic calculation at the level of the cropping or livestock system (subsystems of the production system), for which it is often possible to count separately the quantity of work carried out; (2) the notion of *opportunity cost*, and therefore that of the choice made by farmers to dedicate an hour's or a day's work to one competing activity or another; and finally (3) issues relating to the management of the farmwork planning and the combination of additional activities.

This combination of activities requesting the family to work at different times, leads to increasing global labour productivity, even if daily labour productivity does not increase or even drops.[1] While in industrialised countries, increases in productivity have been determined especially by investment (particularly during the contemporary agricultural revolution), increases in global productivity (much lower) recorded in agrarian systems with very low capital, have been made possible through the progressive filling of farmers' work plan. Maintaining and increasing diversity internal to these systems were at the root of the maintenance or increase in productivity, despite the sometimes complete lack of access to capital. Consequently, what is possible in terms of productivity increase goes beyond capital accumulation, and includes everything that can play on the distribution of work during the year: diversity of productions and spreading of labour peaks, diversification of activities, etc.

For a long time, economists and agricultural economists have had a clear definition and made rigorous use of the word and the concept: a relation between value added during a productive process and a production factor that, in the case of agricultural production, can be the capital, the labour or the land.

Despite the relevance of this notion and its very high fertility in economics, the term experienced a certain dilution once it came to pass into everyday language. Under "productivity", the most used dictionaries systematically refer the reader to the word "yield" and vice-versa, thereby maintaining the idea of the interchangeable nature of these two words.

[1] That is why the fact that marginal labour productivity drops does not prevent global productivity from increasing, a fundamental characteristic of farming economy that had already been suggested in his days by Chayanov (1966) (labour "overintensification").

This is also the case within the agronomists' community, where the term "productivity" is almost always employed to refer to "yield". Yet, until the middle of the twentieth century, when agronomists – acting as economists – started looking into the accounts of the farm, they were talking about "cultivation costs", the "cost price" of cultivation and its "return" (i.e. what it yielded), but not about "productivity", where this term was not used either to mean 'yield'. In fact, in the illustrated encyclopaedia of the *Larousse Agricole* published in 1921, the term is not mentioned; it was still unknown by agronomists.

It was only after the Second World War it seems, that agronomists took over the term "productivity", although most often to refer to "yield". As such, the *Larousse Agricole* published in 1981 proposed the following definition: "In agronomy, production capacity of a species or a variety in a given environment when the optimum cropping conditions are gathered, in other words, maximum yield of a species or a variety in a determined geographic area (. . .) Currently, the productivity of the best varieties of winter wheat is higher than 70 q of dry grains per hectare in the Paris Basin" (Clément 1981). Productivity then is equivalent to "potential yield".

This borrowing by agronomists resulted in too many different things being referred to by the same term, leading to incomprehension and confusion. This confusion took on different forms. The first consisted in using "productivity" to either refer to a simple yield, i.e. gross production related to the surface used or to livestock (to refer for example to a dairy yield), or to a production potential related to the area (this is the sense which is suggested by the *Larousse Agricole* and used to refer to improved, "more productive", "high productivity" varieties or races). The second consisted in including indistinctly in the numerator of this relation, either a gross production (of wheat, milk etc.) or a gross production decreased by the goods and services destroyed to obtain this production, i.e. value added. The third consisted in including indistinctly in the denominator an area (in hectares), capital (measured in monetary unit) or labour (measured in hours or days of work) without the user ever making the effort of specifying which productivity was concerned.

Were it not for the excessive number of times "productivity" served as vector or slogan in the speeches of development agents and political leaders, only purists would have been bothered by this abundance of meaning. The confusion created between yield and productivity illustrates very well, just as it accompanies and serves, the technicist and productivist abuses of contemporary agricultural revolution. By confusing increase in productivity with yield development, it was easy to forget sometimes that an increase in the efficiency of the production process could not take place without controlling the costs, and that it was more the progression of value added, rather than yield alone, which was decisive. The greater part of the agronomic research apparatus and its counterpart in higher agricultural education, had their attention focussed on the progression of the number of quintals per hectare . . . this way of looking at things certainly had an impact during the first oil crisis, and then from the point of view of sustainable development . . .

Moreover, by making of gross production per hectare (i.e. yield) the gold standard of agricultural development instead of productivity (as defined by economists), this confusion resulted in considering work efficiency (labour productivity) and its role in the modernisation of agriculture to be of second importance. Yet it was the unequal progression of labour productivity that determined, more so still than yields, the evolution of agriculture around the world, as shown by Dumont as early as 1954.

Finally, this confusion was not without consequence on the quality of relations established between agronomists and farmers, too often bringing about incomprehension and distrust. While agronomists saw in the increase of "their" productivity (i.e. yield) the spearhead of their modernising action, thereby popularising high "productivity" varieties and races, artificial fertilisers and cultivation tools, farmers assessed the extra work and costs involved and, in this sense, the risk of a decrease in productivity (as defined by farmers and economists alike). This misunderstanding between "technical" ideal and farming rationality which was widely underlain by this semantic confusion, was brilliantly denounced by Pélissier (1979) as regards African farming communities upon which intensification was to be imposed, in a context where the rare factor was not the land but the workforce.

Distribution of VA and Farm Income

As to farm income, the definition adopted by us is that of income resulting from the distribution of value added and from potential transfers made by the community (subsidy). It is equal to the difference between the net value added and all the redistributions reflecting the conditions of access to resources mobilised in the production process (ground rent, remuneration of outside workforce, interests on borrowed capital, taxes on land and products), in addition to subsidies.

Another aspect of economic farm income, which is particularly important in regions and countries where a non-negligible portion of the production is consumed directly by the farmer and his family, is that it is calculated by integrating the overall home consumption, which does represents a share of the value produced per production unit. This income is distinct from *cash* income, although these two results are very frequently confused in the so-called "specialised" literature. Indeed, selling a coffee crop, for example, increases the farmers' cash income but not necessarily the total income. It all depends on the maintenance of the level of satisfaction of the family in food-producing products, and of the utilisation of the cash income and its real buying power. If the cash income is used to buy food-producing products which are no longer produced on the farm, then this "income increase" is a delusion and conceals a loss in the farm's autonomy. In many countries, improving home consumption (diversification and better balance of rations), reducing incurred risks, maintaining and improving the fertility potential of the different farm plots, and increasing farmers' buying power (and not just their "cash income"), cannot be dissociated to make a reproducible production unit of the farm.

But farm income, as previously defined, is not always calculated in this way. With the development, in France for example, of Management Centres or Rural Economic Centres (CER), and the significant increase in the proportion of farms being charged on the real basis, economic criteria (value added, income, productivity etc.) have been decreasingly used and are progressively being relegated to the background, in favour of management, accounting and fiscal criteria in particular: Earnings Before Interest Taxes and Amortization (EBITA) and accounting income. This evolution went hand in hand with the generalisation of personal consultations to individual farmers developed by these institutions. But the GOS, calculated by taking into account the subsidies received but not deducting the depreciation expense, does not come near the value added or farm income; while generally the accounting income assesses only part of the *economic income* cleared by a production system. Moreover, the fact that economists working on agriculture might be tempted to turn to accounting data (considerable sources of information already collected and relatively easy to access) led, in the process, to the quasi-total eviction, as a potential information base, of all farms without book-keeping, usually the most modest farms which, in France, represented the wide majority up until a few decades ago.

Yet this important issue in French agricultural economics research had been pointed out, as early as 1985, in a summary report on the works of INRA's Department of Economics and Rural Sociology. The report talked about the concept of income: "The lack of clarity on knowledge about income in agriculture is, in our mind, found in two main domains: the complete lack of information on certain situations, [and] the confusion being maintained on the concepts and measures used(...). As such, the quasi-totality of those who are in the most unfavourable income conditions are absent from any information on incomes" (Viallon 1985).

The economic approach to production systems differs from the accounting approach in that the latter aims mainly at offering personalised advice to farmers. Consequently, a return to strictly economic criteria becomes essential, as does a certain distance from strictly accounting criteria.

While value added and productivity measure the intrinsic economic efficiency of the production system as value creation process, it is the farm income which is in a position to express the share of value added (potentially increased by the subsidies received), enabling the farmer to support his family and, if possible, to invest so as to increase his capital and, in the end, the productivity of his farm (Fig. 10.1). In family farming, this is the criterion that will best inform about the future of the farm and its capacity to develop.

The conditions of access to production factors, mostly determined by farmers' social relations, influence the conditions for sharing out value added. In family farming, the biggest share of value added produced, is generally allocated to the remuneration of the family workforce, in the form of income, except in situations where land access conditions impose a heavy ground rent (share-cropping) or where access to capital, via all sorts of contractual arrangements (reverse tenancy in particular), drastically reduces the share of value added coming to the farmer. In farming business, once the wage bill and social security contributions have been deducted, and once potential ground rents as well as taxes and levies have

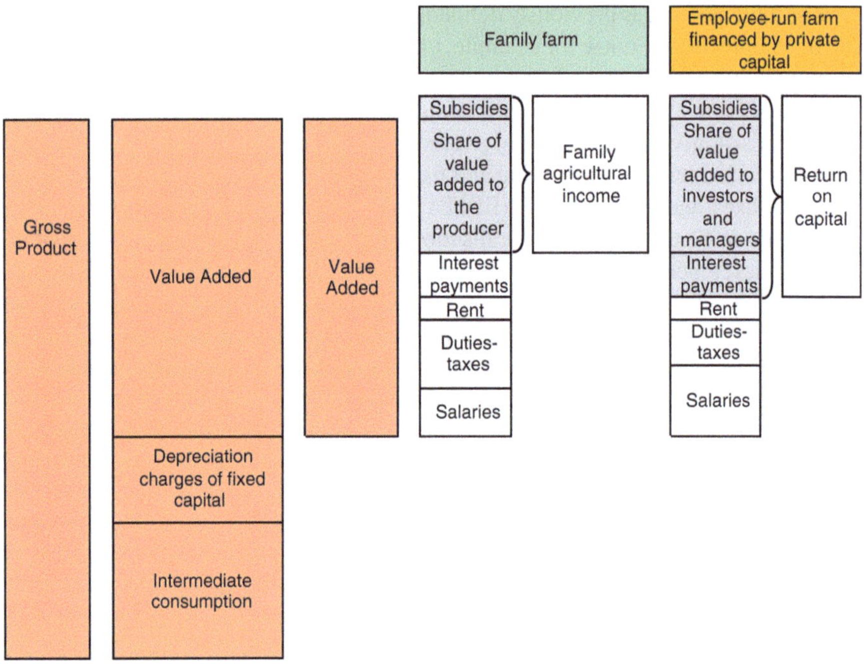

Fig. 10.1 Value added, distribution and farm income

been paid, what remains of the value added produced constitutes the remuneration of the capital invested in the production process. In this case, profitability is more important, i.e. the capacity of the business to remunerate invested capitals, measured via the capital enrichment rate (or profit rate) and the Internal Rate of Return (IRR). Insofar as the economic principles underlying family- and capitalist-type farms are not the same, the farm income and profitability of the invested capital are not very comparable. Moreover, with these criteria, it is not possible to compare the technical and economic efficiency of the different farm types. Yet, the recent development of large-scale farming businesses, particularly within the framework of the ongoing land grabbing phenomenon in the countries of the South and the former Soviet Union by foreign investors, reinforces the importance and the need to conduct such comparisons (Box 10.1). Value added and factor productivity criteria, those governing its distribution and therefore labour and capital remuneration, those making it possible to measure job and income creation, prove to be particularly interesting for comparing the different institutional forms of farming (family farming and agribusiness in particular) sharing the same productive space. Moreover, using these criteria brings up other ratios, and compares conditions for the creation of value added: the share of fossil energy in value added, the weight of chemical inputs in the creation of value added, the respective portion of annual consumption of fixed assets, value added created per m^3 of water consumed (irrigation), among others.

Compared Economic Results: The Ukrainian Example

A particularly meaningful example of the fertility of this approach is given with the study of ongoing dynamics in Ukrainian farming. This farming is characterised by large-scale farms (several thousands of hectares), a reduced salaried workforce and powerful equipment. It is specialised in crop productions which are intended partly for the international market. It is thanks to these production structures that the cereal potential of the rich *chernozems* of this region could open up more easily to the rest of the world, and that Ukraine could be on her way to take her place among the first cereal producers of the world. But in reality, this agriculture is dual in nature. Indeed, the region is also made up of much smaller family farms, among which one finds, in addition to a small number of farms of a few dozen hectares, a multitude of micro-farms (between four and five million nationwide) reduced to farming a patch of less than 1 ha, as under the Soviet era.

Agricultural contractors had to turn to millions of former labourers from *kolkhozy* and *sovkhozy*, to rent the land portions – small in size but put together into very large blocks – that had been distributed during the agrarian reform in the 1990s. Rents were then most often paid in kind (grain and forage in particular) to feed the livestock units of the lessors. Although the authorities do not look upon this small and very small family farm sector as being able to pass off as a fully-fledged economic sector, today it seems to supply an important share of the national production (milk, pork, poultry, potatoes and market garden crops). Consequently, and considering the close organic relations (access to land, flow of labour, transfers of biomass instead of paying ground rents) linking these two facets of Ukrainian farming, how can we apprehend its operation and, a fortiori, anticipate its evolution without analysing *jointly* the different production systems, as contrasted as they are, within the same village or district (see Box 10.1)?

Box 10.1: Performance of Production Systems of Ukrainian Farming Compared Approach at the Level of the Agricultural District (Jaubertie et al. 2010)

An extract from the results obtained in the Province of Jytomyr is presented as an example in the table hereunder, where three indicators are used: the net value added per hectare (NVA/ha) or "land productivity", the net value added per labourer (NVA/labourer), and the number of jobs occupied for 100 ha (jobs/100 ha). The price system used was that which was in force during the campaign of 2008–2009, i.e. that which prevailed once the fever of 2007–2008 dropped.

(continued)

Box 10.1 (continued)

Farm types	NVA/ha Grivna *(Euro)*[a]	NVA/ labourer Grivna *(Euro)*[a]	Jobs/ 100 ha
1/ Farms with mixed crop-livestock farming stemming directly from former Soviet structures: 2,000 ha of barley, wheat, oats, colza, soya and sunflower, a little maize and temporary grassland, 100 dairy cows at 2,200 l, 80 labourers, miscellaneous and partly used equipment	2,500 *(230)*	60,000 *(5,600)*	4
2/ Farms specialised in annual crops: 500–3,000 ha of barley, wheat, oats, colza, soya and sunflower, reduced tillage and precision seeder, high capacity, new and imported equipment, economic results given for 1,000 ha (11 labourers)	1,600 *(150)*	142,000 *(13,000)*	1
3/ Agroholding farm specialised in annual crops of 5,000–30,000 ha. Result for one of the agroholding farms: 5,000 ha of barley, wheat, oats, colza, soya and sunflower, new and imported equipment fleet, 33 labourers	1,800 *(170)*	270,000 *(25,000)*	0.7
4/ Small owner-occupied family farms of 10–30 ha (a land portion of 4 ha + reserve land), a secondhand 40-horse power tractor, cereals + forage crops + vegetable garden, 4 dairy cows at 4,000 l, milk sold at the market, + 2 pigs + poultry, 3 labourers, economic results given for 15 ha	4,200 *(400)*	21,000 *(1,960)*	20
5/ Micro-farms with mixed crop-livestock farming in manual labour (farms belonging to the population): 1 ha of which a 35-are patch of land and 75 ares of the reserve lands rented out + 0.6 ha of collective grazing/cow (+1 land portion of 4 ha rented out), 1 dairy cow at 4,000 l (milk sold at the factory), 2 pigs + poultry, 2 labourers	10,000[b] *(950)*	9,000 *(840)*	110
6/ Micro-farms with mixed crop-livestock farming with animal traction (farms belonging to the population): 2 ha of which a 50-are patch of land + 1.5 ha rented out on the reserve lands + 0.6 ha of collective grazing/cow (+1 land portion of 4 ha rented out), 2 dairy cows at 4,000 l (milk sold to the factory), 2 pigs + poultry, 2 labourers	6,000 *(570)*[b]	10,700 *(1,000)*	55

[a]2009 Euro/Grivna exchange rate: 10.5
[b]This calculation takes into account the area equivalent to the share of the ground rent paid in kind in the form of grains intended for animal feeding, i.e. around 0.2 ha of surface area

(continued)

Box 10.1 (continued)

A comparison of the results recorded in the above table brings out the following results:

- The NVA/ha, or land productivity, is in the region of 140–230 Euro/ha in the large farming structures, in the region of 400 Euro/ha in small family farms with mixed crop-livestock farming and cultivating directly the few land portions, and in the region of 570–950 Euro/ha in the "farms belonging to the population", these proving to be three to five times more productive, per unit area, than large-scale farming businesses, and this although they often develop lands with less agricultural capabilities (grazingland);
- As far as labour productivity is concerned, the hierarchy of the results is obviously reversed, with the means of production (manual tools, animal traction or motorised equipment) resulting in high contrasts. Also, NVA/labourer is in the region of 950 Euros in farms belonging to the population and in the region of 1,960 Euro/labourer in those farming directly their land portions. It is six times higher in structures inherited from *kolkhozy* and *sovkhozy* (around 5,600 Euros) and is established between 13,000 and 25,000 Euro/labourer in farms specialised in annual crops and with new equipment;
- As regards jobs created or maintained per unit area, we find that the best equipped businesses create few jobs (1 job/150 ha) when the large-scale farming businesses stemming from former collective structures, having kept livestock productions, employ six times more labourers per unit area (4 jobs/100 ha). Farms belonging to the population support far more people per unit area, with the number of jobs reaching 60–110/100 ha.

While they are not very efficient as regards job and wealth creation per unit area, farming businesses are of course the farm types where labour productivity is the highest. But the level reached by this last performance criteria (20,000–25,000 Euros per labourer in the agroholdings under study) is extremely sensitive to international grain prices and those of artificial fertilisers and fossil energy. Despite undeniable comparative advantages as far as plot structures and pedological conditions are concerned (in the Tchernozium regions), relative low yields (a maximum of 40–45 quintals/ha of wheat and winter barley) and their irregularity (many climatic uncertainties) limit the agronomic and economic efficiency of these systems.

On the other hand, these businesses appear to be extremely profitable financially-speaking, which explains their recent development. Profitability ratios higher than 10 % or even 20 % can be expected. Two factors explain this high profitability: low rents (12–25 Euros/ha/year only) and low salaries (e.g. a tractor driver/mechanic costs his employer barely more than

(continued)

Box 10.1 (continued)
200–300 Euros/month). Profitability in these farming businesses can therefore be explained by the conditions underlying value added distribution, not by its level.

In the Ukrainian example developed in Box 10.1, the choice of adopted economic criteria throws a different light on the relative "performances" of the different forms of production. By measuring job creation (or maintenance) within the framework of different production systems, one can appreciate the efficiency of each system in this domain, which is particularly sensitive in a country with massive unemployment and immigration.

10.3 The Farmer as Homo-Oeconomicus

Homo-Oeconomicus

In comparative agriculture, it is customary to postulate that farmers, wherever they are in the world, "have good reasons to do what they do" and that, as a result, one must endeavour to find these reasons, especially if the diagnostic carried out must lead to the elaboration of suggestions for the modification of practices. Is a farmer's behaviour "rational" under any circumstance? Answering yes to this question, at least for the purpose of a working hypothesis, would amount to making the basic principle of neoclassic theory ours, i.e. the universal principle of rationality. In reality, things are not quite set out on these terms.

In our approach which advocates a return to field work and surveys, an approach claimed as being *subversive* since it goes against "knowledge" from above, recognising equal rationality from what comes from below really means something.[2] Whether concerning farming communities in the third world which have been denied any form of rationality for too long and are still too often today, or concerning French farmers described as "traditional farmers", now that *rationality is back*, we can extend the field of analysis and open up a domain of knowledge that has not been sufficiently explored (*supra*).

[2]Jean Pierre Deffontaines was one of the first to "dare" make of farming practices a worthy object of research for a research organisation such as INRA, not without difficulty... This approach is related in J. P. Deffontaines, E. landais and M. Benoît: "Les pratiques des agriculteurs, point de vue sur un courant nouveau de la recherche agronomique", (1988).

However, applying the "principle of rationality" to farmers, wherever they are and irrespective of their means of production, cannot be achieved by merely transposing this basic principle of neoclassic theory to the agricultural sector. On the contrary, this principle encompasses a complete reversal of the way things are viewed, in relation to a situation where farmers' decisions and choices would be mostly dictated by custom, routine and tradition, all these explanations being usually considered as artefacts by *Homo oeconomicus* theorists for whom these "archaic residues" would only represent imperfections of the market. Postulating that farmers "have good reasons to do what they do", and therefore seeking in the sequence of their actions and decisions a *rationality*, amounts to extending – by transforming it considerably – the principle of rationality well beyond the narrow circle of rationality, to include the meaning of market-regulated economic agents.

While farmers are rational beings and usually take decisions in conformity with their interests, within of course the material, human and cognitive means to which they have access, nothing indicates on the other hand that all have the same interests, and even less that the maximisation of their production or their income always takes precedence in relation to other interests, as shown by Marc Dufumier (1985). Moreover, farmers take rational decisions not only in relation to the panel of production factors regulated by the market and to which they have access, but also in relation to the conditions, rules and *historical* institutions of access to these factors, and in a perspective of *plural optimisation* (quantity and quality of home consumption, security, increase in cash income, maintenance of long term fertility, etc.), in short, a rationality *situated* in a given historical, social and cognitive context. While the decisions taken by farmers are usually fairly true to their various interests, with the economic calculation often contributing to understanding such decisions, individual choices can only be made according to possibilities offered by a given production system (*supra*).

For all that, while historically developed processes and institutions weigh heavily on the choices made by many farmers across the world, thereby justifying their suspicion towards figures and an "all-economic" approach, this position must not lead to the rejection of quantification under any form, on the contrary. As long as data collection is conducted during meticulous field work by the actual researcher, is guided by clear concepts and on the basis of a judgment sampling, economic calculation can prove to be extremely efficient in explaining the diversity of situations and trajectories, in highlighting the real opportunity costs allocated by producers to the means of production to which they have access and to their workforce, as well as in showing the real comparative advantages – or disadvantages – at their disposal in international competition. That is why it is unfortunate that such economic quantification efforts are sometimes deemed pointless or unachievable (or are carefully avoided), in the name of the rejection of excessive *economiscism*, a rejection which is sometimes called upon in the name of an institutionalism which is somewhat attaching too much importance to the "cultural" factors or "traditional" institutions that would have farmers losing their economic rationality.

Cultural Factor and Tradition

If this issue is addressed here, concerning the economic approach in comparative agriculture, it is precisely because recognising this economic rationality, in the wide sense as seen already, would leave little place to "cultural-type" explanations in understanding agricultural practices. But what meaning and what place should be given to these "cultural factors"?

For a long time, the "cultural factor" or "mentalities" were used as an alibi to negate farmers' knowledge and know-how, or at least to reject their practices in the "traditional", "archaic" or folkloric domain. This was true, almost everywhere in dominated societies during the entire colonial period; it is still the case in most development institutions and projects in the countries of the South, where "mentalities" are called upon to try to explain the lack of enthusiasm, not to say unconcealed hostility, of the populations towards the extension of one "modern" advisory package or another; and this seems to also be the case in a country like France, if we are to believe the number of farmer "typologies" based on categories such as "those who go forward", "the followers" and the "traditional farmers".

Earlier on I pointed out, concerning the virtues of field surveys, how many practices and behaviours remained misunderstood by agronomists, and a fortiori by economists, because they were rejected as being irrational before being observed conscientiously. Isn't "explaining the unexplained through the inexplicable", according to the expression of Jean Pierre Olivier de Sardan (1995), the refuge of the development researcher or practitioner faced with a reality he does not understand? Such explanations, via cultural factors or the irrational, have also been brandished everywhere around the world to enslave "naturally lazy" dominated groups or "*flojos*".[3]

Sometimes still, and more recently, the reintroduction of the "cultural", in reaction to an *economiscism* that would be the cure-all of Western thought, has also led to a radical anti-development, in the name of cultural identity and, increasingly often, in the name of a radical ecology sometimes extending conservationism to actual rural societies. Similarly, "some praise the wonders and pertinence of 'traditional' practices, as well as the ancestral and popular knowledge that was ignored and destroyed by the sudden imposition of industrial techniques. Invoking tradition and ancestors makes these practices and knowledge go back to the dawn of time(. . .), which spares us having to consider today's practitioners as the producers of their own knowledge and practices, a display of condescension similar to that displayed when scientism claims exclusivity" (Darré 1999: 92).[4]

In fact, it is especially when cultural factors and traditions (it does not matter whether they are despised or, on the contrary, idealised) are considered unchanging

[3]This Spanish term is still very much in current usage and is used by certain Latin-American agronomists to refer to Indians . . .

[4]An example of an almost mythical idealisation of ancestral practices is given today by certain indigenist movements from Andean countries.

and are invoked abusively to explain one practice or another, that their usage appears suspicious and unproductive. Putting tradition forward immoderately is to consider that farmers "did not and do not produce the knowledge which directs their practices; they are not supposed to have resources on the subject other than *what came from elsewhere* or *what is already here*" (idem, p. 93). In fact, tradition does not exist as such, or one must then talk about a *set of practices on the move* as suggested by J. P. Darré (idem).

If invoking "cultural" factors amounts to evading everything that has to do with "technical" or "material" issues, or is an expression of discontent about "popular materialism", then it is really about escaping or realising one's failure as far as understanding reality is concerned. If, on the other hand, we consider that practices express part of the relations men forge with their environment, and that these relations are in fact part of their "culture", which no one will contest,[5] then the *material culture* of people becomes again the focus of analysis. The research conducted in comparative agriculture, very often leads to rediscovering this culture which is yet buried, forgotten or repressed in view of the value judgments which agronomists pass too often.

Reversing this tendency and refusing at first this type of "cultural" explanations, abstaining from plunging into this easiness, being a priori wary of cultural explanations before exploring other possible sources of comprehension, this is what the first reaction should be, somehow, for any approach in comparative agriculture. As such, would culture be "what remains after everything else has been envisaged"[6]? Not either, because it is not about turning to the cultural domain as "a last resort" (which in fact depends highly on the extent to which research was studied further...), but rather about seeing the extent to which and why "ways of being" and "ways of doing" can support each other or simply be compatible.[7] Discovering that "ways of being" and "ways of doing" correspond or finding that practices "translate" representations exactly, does not mean that the cultural explanation can suffice but, rather, that culture contains material bases. In this regard, I have shown in the case of Burundi that the whole ideology of the sacred in the former kingdom, recalled every year on the occasion of agrarian rites, came to reinforce in collective representations the new farmwork planning – the economic pillar of the kingdom – with two harvests per year (generalised after the introduction of American plants in the eighteenth century). In this sense, was not the rehabilitation of practices also that of culture?

Moreover, while the technico-economic course of action adopted by farmers is an individual phenomenon, for some, it falls within the competence of psychology.

---

[5]Sigaut goes even further: "The environment is not given in advance. It depends on what people know and what they do, i.e. their culture. The environment of a society forms an integral part of the culture of that society(...). What is pertinent in nature is part of culture..." (Sigaut 1981: 291).

[6]This expression is from Pierre Dumolard (1981).

[7]Such conformity between ways of doing and ways of thinking has been studied in the case of the central Andes, as regards collective rotations (Morlon 1992: 103–104).

This is for example what inspires some to suggest a taste for danger, and others to adopt an overcautious nature in explaining diversity. As a result, many farmers' typologies have been elaborated, separating "those who forge ahead" from "those who follow traditions" with, in the middle, the "followers". Concerning these classifications, Jean Pierre Darré showed how relationships of domination between "those who know" and "those who do", are such that farmers classify themselves according to these predetermined categories[8] (Darré 1999). There is no doubt that some farmers are more prone to "innovate" than others, because it is in their interest and they can afford it. But seeking in the psychology of one individual or another, a more or less marked propensity for risk taking, is not in our intention nor is it in our jurisdiction.[9] Finally, it is not very likely that this "quality" can be isolated from history and from the accumulation methods, which are peculiar to the trajectory of the production unit in question.

[8]See also the paradigm of "diffusion studies" presented by J. P. Olivier de Sardan and for which farmers could be classified into five categories, depending on their attitude towards innovation: the pioneers, the innovators, the early majority, the late majority and the latecomers.... (op. cit. 1995: 82).

[9]Jean Pierre Darré shows the limits of these "psychologising descriptions": "giving a place to what, in explaining actions, is outside reason (...) leads to two consequences. The first, is that I must consider that people do not know why they act the way they do, and the second is that I must question everything they tell me and seek the hidden face behind the mask, ...the 'real reasons' behind their reasons" (1999: 77).

Chapter 11
Comparative Agriculture and Evaluation

The assessment of projects, development programmes or agricultural policy measures, relies on a simple principle: measuring a differential between two situations, that resulting from the establishment of the project/programme/policy on the one hand, and that which would have prevailed should the project not have been established on the other, as recalled by most works and manuals dedicated to project assessment methods (Bridier and Michailof 1980; Casley and Lury 1982; Gittinger 1972; Dufumier 1996; Baker 2000). Thus, highlighting the direct and indirect effects attributable to a project, can only be tackled by recreating this differential between, on the one hand, a scenario “with project” (over the duration corresponding to the estimated “duration of the functional life” of realised investments) and, on the other, a scenario “without project”, also referred to as “counterfactual situation”.

The main difficulty consists in isolating the effects of the project or policy in question, from evolutions already in progress and that would not be attributable to this project or policy. The differential alone (measured in terms of impact on one agent category or another, or in terms of advantages and costs for the community), represents the effect of the project on reality and measures its real impact. Public interventions, as regards agricultural development, are always in line with an agrarian dynamics that also responds to other forces and is only partially influenced, modified, slowed down or, on the contrary, accelerated by these interventions. This dynamics is what needs to be perceived beforehand to be able to ask the right questions, where the impact of development projects is concerned. The exactness of the conclusions put forward will depend mainly on understanding the old and recent transformations of the agrarian system and, in this sense, the development paths identified for each category of producers. This is where the methods and the know-how of comparative agriculture can prove very efficient, and can lead to measuring as close as possible to reality, the true impact of an agricultural policy project or measure.

H. Cochet, *Comparative Agriculture*, DOI 10.1007/978-94-017-9828-0_11

11.1 Systemic Impact Assessment[1]

As obvious as it is, yet the idea of highlighting a differential *[with – without]* project is not always implemented, far from it, especially because the reconstruction of the scenario *without* project encounters many difficulties and relies too often on a few of the assessor's a priori or subjective choices. How would the situation of the "target group" have evolved in the absence of a project? Among the changes observed, which ones are effectively attributable in whole or in part to the project? Which ones, on the contrary, would have occurred in any case if the project had not been established? And on which basis can this issue be addressed? In front of such difficulties, many assessors make do with comparing the situation they can actually observe and measure (the situation "*with* project") with the initial situation, *before* the project, as long as the initial situation has been properly analysed.

Using the "initial situation" instead of a counterfactual scenario (situation without project) to be rebuilt, in fact relies on an implicit hypothesis pregnant with meaning, that of rural societies failing to act, and being unable to transform and evolve outside the project or some exogenous intervention. Although mostly invalidated by facts, as shown by the many works conducted worldwide on this issue, many "developers" and "assessors" still too often make do with this excessively reassuring idea that everything would have remained the same and constant, if an intervention in the form of a project had not taken place.

Moreover, the direct and indirect effects of agricultural development projects are often numerous and very difficult to identify and measure. Where agricultural development projects usually propose innovations likely to spread either via the formal channels established by the project, or outside these, one of the hypotheses the most commonly formulated in project documents is that the distribution of their effects should progressively spread to the villages situated beyond their areas of intervention (according to the conception introduced by Rodgers as early as 1962). Without for all that adhering to this simplistic vision – to say the least – of the road to innovation, it is undeniable that agricultural development projects are often translated into positive as well as negative indirect effects, on populations that are a priori unconcerned with the project: resale of inputs distributed by the project, partial adoption of new disseminated techniques, modification of the market prices, etc.

On the other hand, farmers heading up different production units initially evolve very differently over time, whether or not they are concerned with a project. As such, the elaboration of scenarios with and without project cannot, in the agricultural development domain, only limit itself to compare "concerned" and "unconcerned" individuals, in that such averages would not be much significant. The initial diversity of production units must, on the contrary, be identified beforehand, serve as a basis

[1]For more details, we refer the reader to the paper published by us on this issue (Delarue and Cochet 2013). See also Delarue (2007).

for sampling, as already pointed out as far as the construction of typologies is concerned, and give rise to a modelling in terms of production systems.

That is why, only in-depth knowledge of the endogenous and exogenous factors of evolution, and of the possible trajectories of production units, expanding on former dynamics, can lead to identifying with certainty comparable individuals evolving "with" and "without" the project.

For this, the impact assessment of an agricultural development project must be conducted in an agricultural district, which is homogeneous from the point of view of its agro-ecological characteristics and its agrarian dynamics, and for all production systems existing before the project. As such, the study of the agrarian system and its dynamic, and that of the trajectories followed by the different types of production units peculiar to comparative agriculture, represent the foundation stone of evaluation. The impact assessment of an agricultural development project will therefore result from the many back-and-forth required between systemic diagnostics (at different scales of analysis) and evaluation elements.

By placing the systemic approach back in the centre of the approach for the assessment of agricultural development projects, the *systemic* impact evaluation makes it possible to build scenarios "with" and "without" project for the different types of producers present, and to measure the selected impact indicator(s) in terms of differential and over time.

This approach not only leads to *measuring* the real impact of a project, through one or several assessment criteria, but also offers reliable qualitative information likely to direct decision-makers favourably towards a reorientation of the project, or leading to the formulation of new interventions.

As an example, and to limit ourselves to a few shared experiences, in France, with agronomy students during joint field works, we can mention the case of the policy for the protection of the Cotentin Marsh, established in the Regional Park of the same name. The old history of the marsh is well known and its most picturesque aspects are exhibited in the ecomuseums of the region, or at the Park's house. Yet, a historical analysis of the recent dynamics of the region's production systems, revealed that the measures for the protection of the marsh partly ignored the differentiated utilisation methods of the same marsh, by the different types of local farms. This resulted in the establishment of measures aiming at favouring the "extensive" exploitation of the marsh (late cutting date, regrowth prohibition), which only farms with the best equipment could benefit from via contract (where, in addition, such farms could make unlimited use of the maize silage when distributing rations to their dairy cattle). The small dairy farms, for which exploiting the forage resources of the marsh was essential in mostly grassland forage systems, found themselves de facto excluded from the utilisation proposed by the Park, when in fact their practices turned out to be much closer to the Park's objectives.[2]

[2]Study conducted with Sophie Devienne during an introductory class on agricultural development project engineering (INIP), in the winter of 2002. See also Cochet and Devienne (2002) concerning the experience of territorial farming contracts (CTE).

Likewise, our thoughts developed in 2004 on the impact of ensuring the conformity of livestock buildings on the future of dairy farms in Haute Amance (*Département* of Haute Marne), based on a prior diagnostic, gave us a glimpse of the consequences of this policy on the acceleration of dairy restructuring. The study highlighted the link between past accumulation dynamics, particularly as regards livestock building investments, and the real possibilities for ensuring their conformity by the different types of farm found today. As such, we were able to show that farms with old livestock buildings situated in villages and using milking buckets or pipeline milkers, could not afford the investment, even with financial aid in the form of subsidies, and were therefore condemned to stop their activity or become associated, under rather unfavourable conditions, with a larger farm. Farms with buildings dating from the 1970s and equipped with pipeline milkers or small milking parlours, were in a better position to invest, but experienced a drop of income which, in turn, weakened their position thereafter. Ensuring conformity caused so many farms, in the short and medium term, to stop their activity so fast, that the issue of land and quota takeover by the remaining farms became a concern to local professional managers and administrative officers.[3]

The constant revisions of the Common Agricultural Policy (CAP) of the EU were all occasions to implement this approach and, at the level of the agricultural district, to bring the differentiated impact of these changes on present production systems to light. The impact of delinking part of the public funds on farms, on the levels of production and on the location of productions – the great unknown of this CAP reform at the time when it was decided – was studied in a few French agricultural districts. Studying the trajectories followed by one production system or another put the impact of this reform into perspective, particularly as regards the way the least lucrative crop or livestock units were quickly abandoned, which some foresaw as being unavoidable. The establishment in 2010, of measures decided on the occasion of "CAP Health Check", particularly the increase in funds intended for herbage systems, as well as future changes programmed for 2013, could be the subject of similar systemic impact evaluations.

11.2 Economic Assessment of Development Projects, from the General Interest Point of View

The approach and extent of a national or regional community's interest in an agricultural development process, induced or provoked by an agricultural policy project or measure, calls for a change of scale and, at the same time, the use of specific economic tools.

[3]Study conducted with Sophie Devienne an introductory class on agricultural development project engineering (INIP), in the winter of 2004.

While a financial assessment measures the profitability of an operation, such as an investment, from the point of view of the one who consents to this investment, the economic assessment methods of projects and policies tackle this question from the point of view of the community as a whole. By taking into account in the calculation, all the direct *and* indirect effects of a given investment, upstream and downstream in particular and on the competitive sector, by measuring the value of the goods and services consumed or produced at the *shadow prices*, and by assessing all the national resources consumed during the production process at their *opportunity cost*, this calculation brings out the profitability of an investment or a sector for the national community, and no longer just for the investor (Gittinger 1972; Dufumier 1996, op. cit.).

This approach represents a radical change of perspective on the notions of "profitability" or "competitiveness" of public or private investment projects in agriculture, a change which is bound to create surprises, especially when it highlights the fact that the "sustainable" development of a country, under no circumstances, results from the simple additionality of projects or sectors which are yet presented with good cause as financially profitable. As an example, we can mention that of Ecuadorian industries exporting bananas and shrimps, industries for which profitability is always measured in financial terms, the good financial health of agribusiness being supposed to guarantee the economic development of a country, while on the contrary, "non-profitable" farming production would only have a marginal place in the agricultural economy, and would relate more to a social policy.[4] The approach of the same sectors with the tools of economic evaluation shows that the wealth created per unit area, as impressive as it is (e.g. glasshouse floriculture, shrimp production basins and intensive banana plantations[5]), is translated into a much lower profitability for the national community, due in particular to the fact that a number of negative indirect effects are taken into account: high costs in foreign currency of imported inputs and equipment, massive depreciation of production (at "border" prices) via a fiscal accommodation policy, high opportunity costs of land and water resources, negative externalities as regards health, irreversible damage to the environment, etc.

It is also in terms of economic assessment, from the point of view of the general interest, that investment projects run on a large scale by foreign companies, in the countries of the South and the former Soviet Union, should be studied and compared.

While many made themselves heard, to attract the attention of public opinions and authorities on the possible consequences of these projects, where the environment and social issues are concerned, the issue of their *economic* efficiency, particularly from the *general interest point of view*, on the other hand, has been

[4]The existence in Brazil of a double ministry of farming business and peasant farming illustrates perfectly this social/economic duality attributed to the agricultural sector.

[5]See for example the recent works of Pierre Gasselin on Ecuadorian floriculture (2000) and those of Dario Cepeda on the Ecuadorian production of banana for export (2009).

barely addressed, perhaps due to the fact that many do not in any way doubt this efficiency and the potential increase in production of these projects. And yet it is through the results of a real procedure of economic assessment carried out beforehand, that the following questions could be answered: Will the investment project actually be translated into a significant increase in production, compared to the situation "without project", and how can this increase be measured beforehand? Will the investment project be translated into net job creation, compared to the situation "without project"? We then need to verify and measure the net gains of the project compared to a reference situation "without project", an approach that, in a number of cases, leads to question the relevance of such projects for the host country as well as for the whole of humanity (see Box 11.1).

Box 11.1: Towards an Ex Ante Economic Assessment of Large Scale Land Acquisition/Lease in Developing Countries by Foreign Investors

Large scale land acquisition/lease projects in developing countries are very generally introduced by their authors and justified through the following considerations:

1/ It is imperative to significantly increase agricultural (and energetic) production on a worldwide scale, in order to meet the growing needs of humanity (population increase, progressive generalisation of the consumption patterns of the countries of the North, predictable exhaustion of fossil energy sources).

2/ In developing countries, the agricultural sector is not in a position to face these issues; production and productivity stagnate or do not increase rapidly enough, due in particular to a crucial lack of investment and access capacity to modern technologies.

3/ Since neither States nor the concerned populations (i.e. farmers) have the necessary investment capacities, only (public/private) foreign investors are likely to bring in the necessary capitals. In this sense, foreign investors are likely to replace public aid for development, trending downwards, and local farmers.

4/ Subject to extensive, cheap and, in the long run, stable land access, which gave rise to concerns expressed by many observers in terms of *land grabbing*, foreign investors set out to bring in the capital required for increasing agricultural production. Where the workforce required for the production process can be largely recruited on site, such investment projects would be able to create jobs in rural areas.

What to think about the presentation of these projects? And how to make an economic assessment of them from the general interest point of view, since their promoters would like to see them as part of a "win-win" strategy?

(continued)

Box 11.1 (continued)

We need to propose avenues worth exploring in order to progress towards the establishment of a procedure for the assessment beforehand of investment projects that can offer answers to the following questions:

- How to make sure that the investment project will be translated into a considerable increase in production, and how to measure this increase beforehand?
- How to ensure that the investment project will be translated into net job creation and how to measure this job creation beforehand?

We need to check the *net* gains of the project, i.e. that the establishment of the investment is translated into a significant increase in the indicators in question (production, value added, jobs) over the life cycle of the investment considered, compared to a reference situation assessed over the same cycle.

It is therefore imperative to carry out a diagnostic so as to make an inventory of the initial situation (before the project) as regards the use of resources (land, water, workforce and means of production), and to identify an alternative usage for these resources and production factors, in the event that the investment project is not carried out. This diagnostic must lead to the measurement of the real opportunity costs of local resources dedicated to the project (opportunity costs of the land, of the water in case the project includes an irrigation section, and of the workforce), and therefore to the initiation of an assessment in terms of Profits/Costs.

It is extremely frequent that this net increase (in productions, jobs etc.) is considered self-evident insofar as the farming systems established by farmers in developing countries, are often deemed generally unable to produce efficiently and, a fortiori, to increase production in sufficient proportions. This type of stereotype which is too frequently spread, particularly among investors without experience in this domain, and who were only recently converted to the diversification of their business portfolios in the agricultural domain, has led to the systematic underestimation of (1) the production level reached by these "pre-existing" farming systems, (2) the value added created by these systems, all the higher from the point of view of the yields obtained since the levels of inputs used are often low, (3) the resulting land productivity (net value added per unit area), and (4) the evolution capacity of these systems. On the other hand, it also frequently happens that capital contribution is confused with new technology contribution. Yet, many scientific works have shown that the crucial (and indeed real) lack of capital affecting the agricultural sector of developing countries, is not always an expression – far from it – of a shortage of know-how and technologies.

Establishing such an approach for *ex ante* assessments, puts forward the fact that replacing pre-existing production systems with entirely imported

(continued)

Box 11.1 (continued)
farming systems, based on the production of a small number of foodstuffs, according to crop management sequences that are simplified and major consumers of synthetic inputs and fossil energy, is not always translated into a significant increase in the created value added and in the value added per unit area ratio.

We need to recall that most prospective investment projects in developing countries, concern areas benefitting from conditions that are very much conducive to farming (soil fertility, water resources and accessibility), areas that have been occupied for a long time by agricultural societies often characterised by a relatively high population density, and labour intensive production systems. It is for this reason that, in all likelihood, a considerable proportion of the jobs created might end up replacing existing jobs, which would reflect a net job creation clearly lower (or even negative in some cases) than the objectives displayed. Moreover, in case the local workforce is employed within the framework of the project, we need to check that the opportunity cost of this workforce (the value added it would be in a position to produce if it was allocated to alternative uses, particularly in local production systems) is not prohibitive and does not reduce the value added differential created by investment.

Therefore, we need to be careful in relation to certain hasty claims that would see in these investments an open door to the conquest of "virgin" lands (i.e. unexploited and for which the opportunity cost of the land, the water resources and the workforce would be nil). Such situations actually exist but correspond either to a frontier situation (with a significant impact where the environment is concerned), or to the development of lands momentarily left fallow, following disruptions linked to the collapse of the former USSR (examples in Ukraine and Russia). Everywhere else, large scale investments carried out by public or private foreign agents are always translated into new systems replacing pre-existing agrarian systems, where the real progress as regards the production of value added and job creation, must always be shown beforehand.

In the French context, several agricultural development projects have been subjected to a similar assessment, by teams of researchers and assessors establishing the know-how of comparative agriculture on the subject. The case of the cattle and grassland farming systems of the West, pursuing a sustainable farming approach, was assessed by Nadège Garambois and Sophie Devienne (2010). They show that, by comparing the evolution of a group of farmers involved in this approach for 20 years with that of a "control" group, these new production systems are not only profitable for the said farmers (significant cost reduction), but also contribute to regional or national development through the creation of direct and indirect jobs facilitated by them, and through the additional value added created on that scale (by integrating indirect upstream and downstream value added).

Another example is offered by the economic assessment of collective reservoirs, intended for irrigation and open country vegetable production, which were built during the 1980s and the 1990s on the hillsides of the Béarn region. By taking into account all the direct and indirect advantages of irrigation on the concerned agricultural production systems, as well as on the upstream (supplying of producers) and downstream (food-processing industries in particular) industries, and by taking into account all the costs generated by this type of project – investments (construction of dam and network, individual investments of irrigating farmers, etc.) and operation (at the level of all operators concerned) – the specific contribution of irrigation to regional economic development was highlighted, a contribution that justifies a posteriori this type of project, but that also leads to measuring their sensitivity to the opportunity cost attributed to irrigation water (Cochet et al. 2010b).

These tools for calculating economic assessment, particularly the so-called "reference price method", are in no way specific to comparative agriculture. On the other hand, their usage combined with the analysis of the concrete production process, highlights the real opportunity costs of utilised resources (land, capital and labour mobilised in productive processes), and calculates reference prices measuring at best what is advantageous for the community or, on the contrary, the cost of production or consumption of specific goods or services. By linking the "classic" approaches of economic assessment to the approach in terms of the agrarian system of agricultural dynamics, comparative agriculture makes these assessments more pertinent, in that they are more linked to the production processes, their dynamics and differentiation.[6]

[6] As regards the assessment of agricultural development policies and projects, see also Bazin (1999) and Ducourtieux (2001, 2010).

Conclusion

Comparative agriculture is mostly the result of agricultural economists. Because it involves a vast corpus of knowledge, and because its objective goes well beyond the technical process of production, comparative agriculture is above all a social science. The scientific proximity already mentioned with rural geography, history, economics and agricultural technology (although this could have been extended to anthropology, ethnology or sociology among others), gives an idea of the common basis of knowledge needed for this research subject, i.e. agricultural development. As such, comparative agriculture is at the crossroads of the social and life sciences, and attaches the greatest importance to the technical/social interface. The fact of linking, for example, the measure of the economic performances of an agricultural production system to the imperatives of its technical *operation*, or the fact of placing the farming method of an ecosystem back in the trajectory of its historical evolution and in relation to the social relations built up around access to resources, illustrates the depth granted to this interface and the means implemented to apprehend it in all its dimensions.

But comparative agriculture is not made up of successive borrowings or inputs from other disciplines; it could not be a stack of approaches more or less compatible around a common object, i.e. agriculture. It is not a summary discipline either, in the sense where it would be a summary of other disciplines, and it could not claim to be universalist or encompassing. The proximities highlighted above simply show the necessity to call on all these disciplines as needed; juts as they are an invitation to multidisciplinary dialogue.

Today, comparative agriculture is confronted with a triple requirement:

- It must pursue *cognitive* research on agrarian systems, their origins, differentiation and contemporary transformations, as well as on the crises, transitions or agricultural revolutions punctuating the passage from one system to the other. It must accentuate theoretical synthesis and a synthesis on different scales of analysis, and continue to think about improving and adapting the concepts underlying this discipline. It must pursue and increase the comparison of production processes on a worldwide scale so as to (1) better understand each

H. Cochet, *Comparative Agriculture*, DOI 10.1007/978-94-017-9828-0 process, thanks to the comparatist approach which this work has been about all along, (2) compare productivities for the same foodstuff or category of product, and thus anticipate future evolutions, those of production regions on a worldwide scale in particular, (3) identify and measure the consequences of opening these regions for competition on the dynamics of production systems, as well as their economic results, particularly in terms of jobs and incomes, and (4) extend these comparisons to the measurement of the differentiated impacts of agrarian systems and production systems on environmental dynamics, on a regional and global scale.

- It must contribute to shed light on the public debate and the decisions taken by decision-makers, concerning the production processes to be promoted through its comparatist approach to these processes, their trajectories and differentiation, and its capacity to measure their results as regards quantitative and qualitative production, wealth and income creation, job maintenance or creation, and also from the point of view of the artificialisation of ecosystems and its consequence on the environment. It must contribute to organise research work which is immediately useful to agricultural development professionals, a *finalised* research which is at the service of development, the formulation of projects and policies and their assessment.
- As regards higher education training, comparative agriculture must contribute to endow development professionals with a solid general knowledge on the different farming processes around the world, and to make them able to apprehend the complex situations in which they are led to intervene. Finally, it must equip them with the scientific humility which any practitioner should have before Humanity's agrarian legacy, with the critical mind which is essential for any scientific approach, and with the creativity required for elaborating socially and technically innovative projects.

While this requirement could come across as being excessive, insofar as the issues raised today on the future of the planet, the ecosystems and the different forms of agriculture in the world, call for global, localised, integrated, specific, systemic and multidisciplinary answers, comparative agriculture seems particularly well placed to tackled such issues, perhaps through this sort of *heterosis* effect resulting from the crossing of approaches, and from the combination of spatial and temporal scales.

As to the historical approach, that which puts the interventions of practitioners and engineers specialised in agricultural development, back in the process on which they try to intervene, it seems more than ever necessary. That is why the historical approach is twice as useful: "it makes the enthusiastic planner feel the strength of the sequences he will have to break if he is to replace them with other sequences of evolution. It suggests to the now more modest planner, that he should make his intervention plans compatible with the almost compelling way things develop".[1] The historicity of agricultural development processes, as emphasised by comparative agriculture, also concerns *long* term sustainable development. This last notion demands that "short term everyday actions be in line with long term intergenerational processes and natural resource evolution" (Boiffin et al. 2004: 6).

In the first part of his career, which can be described as "productivist" with all the caution this usage calls for, Dumont was proposing "to compare in order to improve", to compare in order to develop. One generation after the beginning of the agricultural revolution, he dedicated himself to strongly criticising the type of development that stemmed from it. Today it becomes clear that he had detected the abuses and pernicious effects of this revolution – the "negative externalities" as one would call them today – before anyone else had. Proximity to the land, the distance maintained with global theories and the development models advocated here or there, led very early to point out a trail of negative effects intrinsically generated by agricultural development. These ranged from the massive loss of direct jobs in the agricultural sector and a spectacular increase in unemployment (once the possibilities of transfers towards other economic sectors dried up), to various and sometimes irreversible ecological damages and the unreasonable consumption of non-renewable resources, via the increased differentiation and impoverishment of large chunks of the farming community worldwide, among others.

By discovering the chaotic side of this type of development, comparative agriculture turned increasingly to a *critical* approach to development. Today still, and despite the emergence of crosscutting issues challenging the barriers between disciplines, some are still disturbed by this methodological criticism. And yet, criticising "development" should no longer be only considered as adopting an anti-establishment position, as was the case during Dumont's time; especially since, today, it partakes of a societal questioning that affects most disciplines. From this point of view, it is possible that comparative agriculture was – although modestly – in the vanguard of these questionings. That is why today it needs to be a driving force behind clear and thorough diagnoses, as well as proposals at the service of long term development, i.e. sustainable development. While critically reflecting on development, including "sustainable" development, comparative agriculture must help to bring out "the technical models, the organisational forms and the regulation mechanisms ensuring the transition towards a sustainable development which is fulfilled in its different dimensions: economic viability, ecological sustainability and social equity" (Boiffin et al. 2004: 7).

[1] Philippe Couty (1981, op. cit.).

References

Aglietta, M. (1981, 13 mai). Crises et transformations sociales. *Problèmes Economiques, 1.723*.

Apollin, F., & Eberhart, C. (1999). *Análisis y diagnóstico de los sistemas de producción en el medio rural, guía metodológica*. Quito: CICDA/RURALTER.

Assidon, E. (2000). *Les théories économiques du développement*. Paris: La découverte & Syros.

Aubron, C. (2005). Individus et collectifs dans l'appropriation des ressources: le cas d'une communauté andine péruvienne. *Autrepart, 34*, 65–84.

Aubron, C. (2006). *Le lait des Andes vaut-il de l'or ? Logiques paysannes et insertion marchande de la production fromagère andine*. Thèse de doctorat en Agriculture Comparée, INAPG (480 p. + annexe).

Aubron, C., Cochet, H., Brunschwig, G., & Moulin, C. H. (2009). Labor and its productivity in Andean dairy farming systems: a comparative approach". *Human Ecology, 37*, 407–419.

Augé-Laribé, M. (1955). *La révolution agricole*. Paris: Ed. Albin Michel.

Badouin, R. (1987). L'analyse économique du système productif en agriculture. *Cahiers des Sciences Humaines, 23*(3–4), 357–375, Paris: ORSTOM.

Bainville, B., & Dufumier, M. (2007). Transformation de l'agriculture et reconfiguration des terroirs au Sud-Mali: une "pression démographique" à relativiser. *Belgéo, 4*, 403–413. Société belge d'études géographiques. Bruxelles; quatrième trimestre 2007.

Baker, J. L. (2000). *Evaluating the impact of development projects on poverty: a handbook for practitioners*. Washington, DC: The World Bank.

Balandier, G. (1971). *Sens et puissance*. Paris: PUF.

Bazin, G. (Rapporteur général) (1999). *La politique de la montagne, rapport d'évaluation* (deux volumes). Rapport de l'instance d'évaluation présidée par Pierre Blondel, La Documentation française, Paris.

Bazin, G., et al. (1976). *L'agriculture du plateau sud des Dômes, passé, présent et perspectives d'avenir*. Theix: INRA.

Beaud, S., & Weber, F. (2003). *Guide de l'enquête de terrain*. Repères: La découverte.

Bergeret, P., & Deffontaines, J. P. (1986). D'une recherche sur un village du Népal à une recherche sur le développement de ce village. Propositions pour une démarche interdisciplinaire. In *Dynamique des systèmes agraires, L'exercice du développement* (pp. 11–39). Paris: éditions de l'ORSTOM, coll. Colloque et Séminaires.

Bernstein, H., & Byres, T. J. (2001). From peasant studies to agrarian change. *Journal of Agrarian Change, 1*(1), *1–56*.

Bertrand, G. (1975). Pour une histoire écologique de la France rurale. *Histoire de la France rurale, t.1*, sous la direction de Georges Duby et Armand Wallon, Editions du Seuil, Paris, pp. 34–113.

H. Cochet, *Comparative Agriculture*, DOI 10.1007/978-94-017-9828-0

Bertrand, G., & Bertrand, C. (1978). Le paysage entre la nature et la société. *Revue géographique des Pyrénées et du Sud-Ouest, avril 1978, tome 49, fascicule 2*. Presses Universitaires Le Mirail-Toulouse.

Biba, G., & Pluvinage, J. (2006, novembre–décembre). La pluriactivité dans l'exploitation agricole, transition ou composante durable de l'organisation de la production en Albanie. *Cahiers d'Agriculture, 15*(6), 535–541.

Blaikie, P. (1985). *The political economy of soil erosion in developing countries*. New York: Pearson Education, Longman Scientific & Technical.

Blanchemanche, P. (1990). *Bâtisseur de paysages*. Paris: Editions de la Maison des Sciences de l'Homme.

Blanc-Pamard, C. (1986). Dialoguer avec le paysage ou comment l'espace écologique est vu et pratiqué par les communautés rurales des hautes terres malgaches. In C. Blanc-Pamard, et al. (Eds.), *Milieux et Paysages, essai sur diverses modalités de connaissance* (pp. 17–35). Paris: Masson.

Blanc-Pamard, C. (1990). Lecture de paysage, une proposition méthodologique. In J. F. Richard (éditeur scientifique), *La dégradation des paysages en Afrique de l'Ouest, points de vue et perspectives de recherches* (pp. 269–280). Dakar: AUPELF/UICN/ORSTOM/ENDA.

Blanc-Pamard, C., & Milleville, P. (1985). Pratiques paysannes, perception du milieu et système agraire. *A travers champs, agronomes et géographes* (pp. 101–138). ORSTOM, coll. Dynamique des systèmes agraires.

Bloch, M. (1931). *Les caractères originaux de l'histoire rurale française*, éd "Belles lettres" Paris-Oslo. (réédition Armand Colin, Paris, 1976).

Bloch, M. (1949). *Apologie pour l'histoire ou le métier d'historien*. Paris: A. Colin.

Boiffin, J., Hubert, B., & Durand, N. (2004, Nov). *Agriculture et Développement Durable, enjeux et questions de recherche*. Paris: INRA.

Bonnamour, J. (1993). *Géographie Rurale, position et méthode*. Paris: Masson.

Boserup, E. (1965). *The conditions of agricultural growth, the economics of agrarian change under population pressure*. London: George Allen & Unwin LTD.

Boulaine, J., & Legros, J. P. (1998). *D'Olivier de Serres à René Dumont, portraits d'agronomes*. Paris: Lavoisier, TEC & DOC.

Bridier, M., & Michailof, S. (1980). *Guide pratique d'analyse de projets; Evaluation et choix des projets d'investissements*. Paris: Economica.

Brossier, J. (1987). Système et système de production, note sur les concepts. *Cahiers des Sciences Humaines, 23*(3–4), 377–390, Paris: ORSTOM.

Brossier, J., & Petit, M. (1977). Pour une typologie des exploitations agricoles fondée sur les projets et les situations des agriculteurs. *Economie Rurale, 122*, 31–40.

Brossier, J., de Bonneval, L., & Landais, E. (Eds.) (1993). *Systems studies in agriculture and rural development*. Paris: INRA.

Brun, V. (2008, Janvier-Mai). Secteur privé et céréaliculture familiale dans le Mexique du libre-échange. Une étude dans les terres basses du sud-Veracruz. *Economie Rurale, 303-304-305*, 90–107.

Brunet, R. (1993). *Les mots de la Géographie, Dictionnaire critique*. Paris: Reclus, la Documentation française.

Capillon, A., & Manichon, H. (1979). Une typologie des trajectoires d'évolution des exploitations agricoles (principes, application au développement agricole régional). *Extrait du procès-verbal de la séance du 10 octobre 1979* (pp. 1168–1178). Académie d'Agriculture de France.

Casley, D. J., & Lury, L. A. (1982). *Monitoring and evaluation of agriculture and rural development projects*. Washington DC: World Bank.

Cepeda, B. D. (2009). *Ces mains qui font le régime. Dynamique et performances agro-économiques des systèmes de production bananiers en Equateur*. Ph.D. dissertation AgroParis Tech, UR Systèmes Agraires et développement Rural, directeur of the tesis H. Cochet, Paris (431 p.).

Chauveau, J. P. (1999). L'étude des dynamiques agraires et la problématique de l'innovation. In Introduction à l'ouvrage collectif: J. P. Chauveau, M. C. Cormier-Salem, & E. Mollard (Eds.),

L'innovation en agriculture, Questions de méthodes et terrains d'observation (pp. 9–31). Paris: IRD Editions, coll. "A travers Champs".

Chauveau, J. P., Cormier-Salem, M. C., & Mollard, E. (1999). *L'innovation en agriculture, Questions de méthodes et terrains d'observation*. Paris: IRD Editions, coll. "A travers Champs".

Chayanov, A. V. (1966). *The theory of peasant economy*. Homewood: R.D. Irwin (first edition in German, 1923).

Cheyroux, B. (2005). *Intensification et diversification de l'agriculture dans la région deltaïque de Damnoen Saduak – Des fruits et des légumes dans le grenier à riz thaïlandais*. Thèse de doctorat en Agriculture Comparée, INAPG. (Médaille de thèse de l'Académie d'Agriculture en 2005).

Cheyroux, B., & Dufumier, M. (2006). The development of a horticultural agrarian system in the Mae Khlong and Chao Phraya deltas (Thailand). *Moussons, 9–10*, 131–155, Marseille.

Chia, E., Dugué, P., & Sakho-Jimbira, S. (2006, novembre-décembre). Les exploitations agricoles sont-elles des institutions ? *Cahiers Agricultures, 15*(6), 498–505.

Cholley, A. (1946). Problèmes de structure agraire et d'économie rurale. *Annales de géographie, 298*, 81–101, *LVè année, avril-juin 1946*.

Chombart de Lauwe, J., & Poitevin, J. (1957). *La gestion des exploitations agricoles*. Paris: Dunod.

Chrétien, J. P. (1993). *Burundi: l'histoire retrouvée*. Paris: Karthala.

Clément, J. M. (1981). *Larousse agricole*. Paris: Larousse.

Cochet, H. (1993). *Des barbelés dans la Sierra, origines et transformations d'un système agraire au Mexique*. Paris: ORSTOM Editions, coll. A travers Champs.

Cochet, H. (1998, mai). *Importance et enjeu d'une recherche approfondie sur l'histoire agraire de l'ancien Ciskei et de ses environs*. Rapport de mission en République d'Afrique du Sud, INA P-G/INRA-SAD.

Cochet, H. (2001). *Crises et révolutions agricoles au Burundi*. Paris: INA P-G/Karthala.

Cochet, H. (2003). A half century of agrarian crisis in Burundi (1890–1945): the incapacity of the colonial administration in managing the agrarian crisis of the late eighteen-hundreds. *African Economic History, 31*, Wisconsin: Madisson.

Cochet, H. (2004). Agrarian dynamics, population growth and resource management: the case of Burundi. *GeoJournal, An International Journal on Human Geography and Environmental Sciences, 60*, 111–122, Kluwer Academic Publishers, The Netherlands.

Cochet, H. (2005). Concurrence déloyale, créativité écrasée: l'agriculture vivrière en crise. *Esprit Août-septembre 2005*, 33–44.

Cochet, H. (2008a, juillet-août). Vers une nouvelle relation entre la terre, le capital et le travail en agriculture. *Etudes foncières, 134*, 24–29.

Cochet, H. (2008b). Recherches en cours sur les transformations contemporaines de l'agriculture éthiopienne. *Annales d'Ethiopie, 2007-2008, XXIII*, 441–469.

Cochet, H., & Devienne, S. (2002, juillet-août). Premières réflexions sur la mise en place des contrats territoriaux d'exploitation dans le département de la Meuse. *Economie Rurale, 270*, 73–83.

Cochet, H., & Devienne, S. (2004, 18–19 Nov). Comprendre l'agriculture d'une région agricole: questions de méthode sur l'analyse en termes de systèmes de production. Colloque SFER *Les systèmes de production agricole: performances évolutions perspectives*, Lille.

Cochet, H., & Devienne, S. (2006, novembre-décembre). Fonctionnement et performances économiques des systèmes de production agricole: une démarche à l'échelle régionale. *Cahiers Agricultures, 15*(6), 578–583.

Cochet, H., Léonard, E., & de Surgy, J. D. (1988). *Paisajes agrarios de Michoacán*. Zamora: El Colegio de Michoacán.

Cochet, H., Brochet, M., Ouattara, Z., & Boussou, V. (2002). *Démarche d'étude des systèmes de production de la région de Korhogo – Koulokakaha – Gbonzoro en Côte d'Ivoire*. Paris: Les Editions du GRET, coll. Agridoc Observer et comprendre un système agraire.

Cochet, H., Devienne, S., & Dufumier, M. (2007, janvier–mars). L'Agriculture Comparée, une discipline de synthèse ? *Economie Rurale, 297–298*, 99–112.

Cochet, H., Léonard, E., & Tallet, B. (2010a). Le métayage d'élevage au Mexique, Colonisation foncière et dynamique d'une institution agraire dans l'histoire contemporaine. *Les Annales de Géographie* (sous presse).

Cochet, H., Ducourtieux, O., Dufumier, M., & Peyre, Y. (2010b). *Quelle contribution de l'irrigation au développement régional ? Evaluation économique d'un projet d'irrigation dans les Coteaux du Béarn, le cas du barrage de Boueilh-Boueilho-Lasque (Pyrénées Atlantiques).* Rapport d'expertise AgroParisTech (140 p + annexes).

Colin, J.-P. (1990). Regard sur l'institutionnalisme américain. *Cah. Sci. Hum., 26*(3), 365–377.

Colin, J. P. (2003). *Figures du métayage, Etude comparée de contrats agraires (Mexique).* Paris: IRD Editions, coll. A travers champs.

Colin, J.-Ph., & Losch, B. (1994). But where on Earth has Mamadou hidden his production function? French Africanist Rural Economics and Institutionalism. In J. Acheson (Ed.), *Anthropology and institutional economics* (pp. 331–363). Lanham: University Press of America.

Conway, G. R. (1984). Agroecosystem analysis. *Agricultural Administration, 20*(1), 31–55.

Couty, P. (1984). La vérité doit être construite. *Cahiers des Sciences Humaines, 20*(1), 5–15.

Couty, Ph. (1981). Le temps, l'histoire et le planificateur, AMIRA n° 32, INSEE-Coopération, 1981 (texte réédité in Couty (Ph.), 1996: *Les apparences intelligibles, une expérience africaine*, éditions Arguments, Paris, pp. 163–170).

Darré, J. P. (1999). *La production de connaissance pour l'action, Argument contre le racisme de l'intelligence.* Paris: MSH/INRA.

de Bonneval, L. (1993). *Systèmes agraires Systèmes de production, Vocabulaire franco-anglais.* Paris: INRA.

de Schlippé, P. (1956a). *Ecocultures d'Afrique*, l'harmattan/Terres et vie, Belgique, titre original: *Schfting cultivation in Africa, The zande System of agriculture.* London: Routledge & Kegan Paul.

de Schlippé, P. (1956b). De l'anthropologie agricole. *Problèmes d'Afrique centrale, III.*

Deffontaines, J. P. (1973). Analyse du paysage et étude régionale des systèmes de production agricole, *Economie Rurale, 98*, 3–13.

Deffontaines, J. P. (1991). Champ in *Histoires de géographes* (pp. 27–42). Mémoires et Documents de géographie, CNRS.

Deffontaines, J. P. (1997). Du paysage comme moyen de connaissance de l'activité agricole à l'activité agricole comme moyen de production du paysage. In *Thème et variations, nouvelles recherches rurales au Sud* (pp. 305–321). ORSTOM/Centre d'Etudes Africaines URA 94, Orstom éditions, coll. Colloques et Séminaires.

Deffontaines, J. P., & Osty, P. L. (1977). Des systèmes de production agricole aux systèmes agraires, Présentation d'une recherche. *L'Espace Géographique, 3*, 195–199, Paris.

Deffontaines, J. P., Landais, E., & Benoît, M. (1988, janvier-mars). Les pratiques des agriculteurs, point de vue sur un courant nouveau de la recherche agronomique. *Etudes Rurales, 108.*

Delarue, J. (2007). *Mise au point d'une méthode d'évaluation systémique d'impact des projets de développement agricole sur le revenu des producteurs, étude de cas en région Kpele (République de Guinée)*, thèse de doctorat en Agriculture Comparée, AgroParisTech, Paris (414 p + annexes)

Delarue, J., & Cochet, H. (2013). Systemic impact evaluation: a methodology for complex agricultural development projects. The case of a contract farming project in guinea. *European Journal of Development Research, 25*(5), 778–796.

Devienne, S. (1997). *Etude-diagnostic de la situation agricole de la section rurale de Mathador, commune de Dondon – Haïti*, INAPG/FAMV, ronéotypé (59 p.).

Devienne, S. (2002). Les leçons de l'agriculture américaine, ou la quintescence de la méthode René Dumont. In M. Dufumier (coor.), *Un agronome dans son siècle, actualité de René Dumonti* (pp. 45–53). Paris: Karthala/INA P-G.

Devienne, S. (2006). Spectacular rice growing intensification in the Red River Delta: half a century of evolution. *Moussons, 9–10.*

Devienne, S., & Wybrecht, B. (2002). Analyser le fonctionnement d'une exploitation, *Mémento de l'agronome* (pp. 345–372). Paris: CIRAD – GRET Ministère des Affaires étrangères.

Devienne, S., Bazin, G., & Charvet, J. P., (2005, janvier-février). Politique agricole et agriculture aux Etats-Unis: évolution et enjeux actuels. *Annales de géographie, 641*, 3–26.

Diamond, J. (2005). *Collapse, how societies chose to fail or succeed*. New York: Penguin Books.

Dollé, V. (1984). Les outils et méthodes du diagnostic sur les systèmes d'élevage. *Les Cahiers de la Recherche Développement, 3–4*, 89–96, Montpellier.

Duby, G., & Wallon, A. (sous la direction de) (1975). *Histoire de la France Rurale* (4 tomes), Paris: Seuil.

Ducourtieux, O. (2001). *Document de projet – PDDP phase 2: Document C (évaluation économique) et D (annexe évaluation économique)*. Phongsaly: PDDP (ronéotypé).

Ducourtieux, O. (2010). *Du riz et des arbres. L'interdiction de l'agriculture d'abattis-brûlis, une constante politique au Laos*. Paris: IRD-KARTHALA.

Ducourtieux, O., et al. (2005). Land policy and farming practices in Laos. *Development and Change, 36*(3), 499–526.

Dufumier, M. (1985, avril). Systèmes de production et développement agricole dans le Tiers Monde. *Les cahiers de la Recherche-Développement, 6*, Montpellier.

Dufumier, M. (1996a, nov). Systèmes agraires et politiques agricoles, *Politiques agricoles/Agricultural policy*. In M. Sébillotte (Ed.), *Recherches-système en agriculture et développement rural* (pp. 926–931). actes du Symposium International, Montpellier.

Dufumier, M. (1996b). *Les projets de développement agricole, Manuel d'expertise*. Paris: CTA – Karthala.

Dufumier, M. (2002a). *Un agronome dans son siècle, actualités de René Dumont*. Paris: Karthala/INA P-G.

Dufumier, M. (2002b). Economie agricole dans le monde et agriculture comparée. In M. Dufumier (Ed.), *Un agronome dans son siècle, actualités de René Dumont* (pp. 61–68). Paris: Karthala/INA P-G.

Dufumier, M. (2004). *Agricultures et paysanneries des Tiers mondes*. Paris: Karthala.

Dufumier, M. (2006, novembre-décembre). Diversité des exploitations agricoles et pluriactivité des agriculteurs dans le Tiers Monde. *Cahiers Agricultures, 15*(6), 584–588.

Dufumier, M. (2007, juillet-septembre). Agriculture Comparée et développement Agricole. *Revue Tiers Monde, 191*, 1–16.

Dufumier, M., & Bergeret, P. (2002). *Analyser la diversité des exploitations agricoles*. In CIRAD – GRET, 2002: Mémento de l'agronome, CIRAD – GRET, Ministère des Affaires étrangères, Paris, pp. 321–344.

Dumolard, P. (1981). Le point de vue culturel en géographie: réactions dermiques et épidermiques. *L'Espace Géographique* géographie, *4*, 295–298.

Dumont, R. (1935). *La culture du riz dans le delta du Tonkin*. Paris: Société d'éditions géographiques, maritimes et coloniales.

Dumont, R. (1949). *Les leçons de l'Agriculture américaine*. Paris: Flammarion.

Dumont, R. (1951). *Voyages en France d'un agronome*. Paris: Editions M.-Th Génin.

Dumont, R. (1952). Agriculture Comparée. article du Larousse Agricole, pp. 903–938.

Dumont, R. (1953). *Titres et travaux*, notice rédigée pour sa présentation au concours de Professeur. Paris (ronéotypé): INA.

Dumont, R. (1954). *Economie agricole dans le monde*. Paris: Dalloz.

Dumont, R. (1957). *Révolution dans les campagnes chinoises*. Paris: Editions du Seuil.

Dumont, R. (1962). *L'Afrique noire est mal partie*. Paris: Editions du Seuil.

Dupré, G. (1991). *Savoirs paysans et développement*. Paris: Karthala-ORSTOM.

Ellis, F. (2000). *Rural livelihoods and diversity in developing*. New York: Oxford University Press.

Fairhead, J., & Leach, M. (1996). *Misreading the African Landscape. Society and ecology in a forest-savanna mosaic*. Cambridge: Cambridge University Press.

FAO. (1999). *Guidelines for agrarian systems diagnosis*. Rome: Sustainable Development department, Rural Development Division, Land Tenure Service.

Fauroux, E. (2003). *Comprendre une société rurale. Une méthode d'enquêtes anthropologique appliquée à l'Ouest malgache* (152 p.). Edition du GRET/IRD, coll. Etudes et travaux.

Fénelon, P. (1970). *Vocabulaire de géographie agraire*. Gap: Publication de la Faculté des lettres et sciences humaines de Tours - 2, Imprimerie Louis-Jean.

Ferraton, N., & Touzard, I. (2009). *Comprendre l'agriculture familiale. Diagnostic des systèmes de production*. Paris: Quae Editions.

Ferraton, N., Cochet, H., & Bainville, S. (2003). *Initiation à une démarche de dialogue, étude des systèmes de production dans deux villages de l'ancienne boucle du cacao (Côte d'Ivoire)*. Paris: Les Editions du GRET, coll. Agridoc "Observer et comprendre un système agraire".

Freguin, S., & Devienne, S. (2006, juillet–septembre). Libéralisation économique et marginalisation de la paysannerie en Haïti: le cas de l'Arcahaie. *Revue Tiers Monde, 187*, 621–642.

Fresco, L. (1984). Comparing anglophone and francophone approaches to Farming Systems Research and Extension. *Farming Systems Support Project Networking Paper, 1*, 36.

Garambois, N. (2011). *Des prairies et des hommes. Les systèmes herbagers économes du bocage poitevin: agroécologie, création de richesse et emploi en élevage bovin*. Ph.D. dissertation. Paris: AgroParisTech.

Garambois, N., & Devienne, S. (2010). Evaluation des systèmes de production innovants inscrits en agriculture durable: le cas des systèmes bovins herbagers du Haut-bocage poitevin. Colloque international *Innovation et Développement Durable, ISDA 2010*, Montpellier 28 juin–1er juillet 2010.

Gasselin, P. (2000). *Le temps des roses: la floriculture et les dynamiques agraires de la région agropolitaine de Quito (Equateur)*. Paris: Thèse de doctorat en Agriculture Comparée et Développement Agricole, INAPG.

Gastellu, J. M. (1979). Mais où sont donc ces unités économiques que nos amis cherchent tant en Afrique ?, *Note de travail, Série: enquêtes et outils statistiques, vol 1, "Le choix de l'unité"* (pp. 99–122). AMIRA.

George, G., & Verger, F. (2000). *Dictionnaire de la Géographie* (1ère édition 1970). Paris: PUF.

Gervais-Lambony, Ph. (2003). Quelques remarques générales sur la comparaison en sciences sociales en général, et en géographie en particulier. In Ph. Gervais Lambony, F. Landy, & S. Oldfield (Eds.), *Espaces arc-en-ciel, Identité et territoires en Afrique du Sud et en Inde* (pp. 29–40). Paris: Géotropiques Université de Paris-X/KARTHALA/IFAS.

Gibon, A., Sibbald, A. R., Flamant, J. C., Lhoste, P., Revilla, R., Rubino, R., & Sorensen, J. T. (1999). Livestock farming systems research in Europe and its potential contribution for managing towards sustainability in livestock farming. *Livestock Production Science, 61*, 121–137.

Gittinger, J. P. (1972). *Economic analysis of agriculture project*. Baltimore, Maryland: Johns Hopkins University Press.

Gourou, P. (1973). *Pour une géographie humaine*. Paris: Flammarion.

Griffon, M. (2006). *Nourrir la planète. Pour une révolution doublement verte*. Paris: Odile Jacob.

Guichaoua, G., & Goussault, Y. (1993). *Sciences sociales et développement*. Paris: Armand Colin.

Guillaud, D. (1993). *L'ombre du mil, un système agropastoral en Aribinda (Burkina Faso), éditions de l'ORSTOM, coll.* Paris: A travers Champs.

Haudricourt, A. G. (1987). *La technologie, science humaine, recherches d'histoire et d'ethnologie des techniques*. Paris: MSH.

Haudricourt, A. G., & J.-Brunhes Delamare, M. (1955). *L'Homme et la charrue à travers le monde*. Paris: Gallimard.

Hervieu, B. (2002). Le livre d'une rupture: Le problème agricole français. Esquisse d'un plan d'orientation et d'équipement. In M. Dufumier (Ed.), *Un agronome dans son siècle, actualités de René Dumont* (pp. 41–44). Paris: Karthala/INA P-G.

Hugon, Ph. (1992). La méso-économie institutionnelle et l'agriculture africaine: le cas de la filière coton. In M. Griffon (éditeur scientifique), *Economie institutionnelle et Agriculture – Institutional Economics and Agriculture* (pp. 193–210). Actes du XIIIème séminaire d'économie rurale, 7, 8, 9 septembre 1992. Montpellier: Michigan State University/CIRAD/Indiana University.

IAASTD (International Assessment of Agricultural Science Knowledge and Technology for Development). (2009). *Agriculture at a crossroad, global report*. Edited by:

Bervely, D. McIntyre (IAASTD, secretariat), Hans, R. Herren (Millenium Institute), Judi Wakhungu (African Centre for Technology Studies), Robert, T. Watson (University of West Anglia). Washington, DC: Island Press.

INRA. (1977). *Pays, Paysan Paysage dans les Vosges du sud, les pratiques agricoles et la transformation de l'espace*, (ouvrage collectif). Paris: INRA édition.

INRA. (1986). *Les collines du Népal central. Ecosystèmes, structures sociales et systèmes agraires*: tome I: *Paysages et sociétés dans les collines du Népal*; tome II: *Milieux et activités dans un village népalais*, (ouvrage collectif dirigé par J.F. Dobremez). Paris: INRA.

Jamin, J. Y., Havard, M., Mbétid-Bessane, E., Djamen, P., Djonnewa, A., Djondang, K., & Leroy, J, (2007). Modélisation de la diversité des exploitations. In M. Gafsi, P. Dugué, J. Y. Jamen, & J. Brossier (coord.), *Exploitations agricoles familiales en Afrique de l'Ouest et du Centre* (pp. 123–153). Versailles: CTA-QUAE.

Jaubertie, C., Pardon, L., Cochet, H., & Levesque, R. (2010). Ukraine: une approche comparée des dynamiques et performances économiques des structures agricoles. *Notes et Etudes Socio-Economiques, 2010*, Ministère de l'Alimentation, de l'Agriculture et de la Pêche, Paris (sous presse).

Jobbé-duval, M. (2005a). *Mil et une recettes de pomme de terre. Dynamiques agraires et territoriales à Altamachi, Cordillère orientale des Andes bolivien*, thèse de doctorat en Agriculture Comparée, INAPG (300 p + annexes)

Jobbé-duval, M. (2005b). Las haciendas de Norte Ayopaya en el Período Republicano: Una forma de control social del territorio. *Búsqueda, 25*, 177–191, Cochabamba, Bolivia.

Johnston, R. J., Gregory, D., Pratt, G., & Watts, M. (2000) *The dictionary of Human Geography* (4th ed.).

Jouve, Ph. (1986). Quelques principes de construction de typologies d'exploitations agricoles suivant différentes situations agraires. In *Les cahiers de la Recherche-Développement* (Vol. 11, pp. 48–56).

Jouve, Ph. (1988, décembre). Quelques réflexions sur la spécificité et l'identification des systèmes agraires. *Les Cahiers de la Recherche Développement, 20*, 5–16.

Jouve, Ph., & Tallec, B. (1994, nov.). Une méthode d'étude des systèmes agraires en Afrique de l'Ouest par l'analyse de la diversité et de la dynamique des agrosystèmes villageois. Communication au *Symposium international sur recherche-systèmee en agriculture et développement rural*. Montpellier, France.

Khon Kaen University. (1987). *Rapid Rural Appraisal*. Proceedings of the 1985 International Conference, Rural Systems Research and Farming Systems Research Projects, Khon Kaen University, Thailand.

Kroll, J. C. (1992). *Les politiques publiques dans le développement de l'agriculture française et européenne*. Paris: HDR Université de Paris X Nanterre.

Lamballe, P., & Castellanet, C. (2003). *La recherche-action en milieu paysan: méthode et outils*. Paris: Les éditions du GRET.

Landais, E. (1992). Principes de modélisation des systèmes d'élevage. *Les Cahiers de la Recherche Développement* (Vol. 32, pp. 82–95). Montpellier.

Landais, E., & Balent, G. (1995). Introduction à l'étude des pratiques d'élevage extensif. In E. Landais (Ed.), *Pratiques d'élevage extensif: identifier, modéliser, évaluer* (pp. 13–36). Paris: INRA, collection Etudes et recherches sur les systèmes agraires et le développement.

Larrère, G. R. (1974). *Considérations générales - et quasiment théoriques – sur les systèmes agraires, point de vue qui en dérive quant à l'articulation des recherches des biologistes et des économistes dans l'ATP environnement (Région des Dômes),* Note, octobre 1974 (ronéotypée).

Latouche, S. (2001, mai). En finir, une fois pour toutes avec le développement. *le Monde Diplomatique*.

Lavigne Delville, Ph., Bouju, J., & Le Roy, E. (2000). *Prendre en compte les enjeux fonciers dans une démarche d'aménagement. Stratégies foncières et bas-fonds au Sahel*. Paris: Editions du GRET, coll. Etudes et travaux.

Le Roy Ladurie, E. (1969). *Les paysans du Languedoc*. Paris: Flammarion.

Le Roy, E., Karsenty, A., & Bertrand, A. (1996). *La sécurisation foncière en Afrique. Pour une gestion viable des ressources renouvelables*. Paris: Karthala.

Lebeau, R. (2000). *Les grands types de structure agraire dans le monde*. Armand Colin (7ème édition mise à jour).

Levy, J., & Lussault, M. (2003). *Dictionnaire de la Géographie et de l'espace des sociétés*. Paris: Belin.

Lhoste, Ph. (1984). Le diagnostic sur le système d'élevage. *Les Cahiers de la Recherche Développement* (Vol. 3–4, pp. 84–88). Montpellier.

Lizet, B., & de Ravignan, F. (1987). *Comprendre un paysage. Guide pratique de recherche*. Paris: INRA.

Malassis, L., Cépède, M., Bergmann, D., & Dumont, R. (1954, janvier). La petite région agricole, contribution à l'étude et à la réorientation de l'économie agricole d'une petite région. *Economie rurale 19*.

Malthus, T. R. (1798). *Essay on the principle of population*. London.

Mayaud, J. L. (1999). *La petite exploitation rurale triomphante, France XIXè siècle*. Paris: Belin.

Mazoyer, M. (1974). *Itinéraire*. Paris: INA P-G (ronéotypé).

Mazoyer, M. (1987). *Dynamique des Systèmes Agraires, Rapport de synthèse présenté au Comité des systèmes agraires*. Paris: Ministère de la Recherche et de la Technologie.

Mazoyer, M. (1989). Agriculture comparée article, dans l'édition 1989 du *Grand Larousse Universel*.

Mazoyer, M., & Roudart, L. (1997a). *Histoire des agricultures du monde, du néolithique à la crise contemporaine*. Paris: Seuil.

Mazoyer, M., & Roudart, L. (1997b). Pourquoi une théorie des systèmes agraires ? *Cahiers Agricultures, 6*, 591–595.

Mazoyer, M., & Roudart, L. (2006). *A history of world agriculture, from the neolithic to the current crisis* (528 p.). London/Sterling: EARTHSCAN.

Mazoyer, M., Roudart, L., & Mayaki, I. A. (2008). Rapport sur le Développement dans le Monde, 2008, Banque Mondiale. L'Agriculture au service du développement. Résumé et commentaires. *Monde en Développement, 143*, 117–136.

McCann, J. C. (1995). *People of the plow. An agricultural history of Ethiopia 1800–1990*. USA: Madison, The University of Wisconsin Press.

McCann, J. C. (2005). *Maize and grace. Africa's encounter with a new world crop 1500–2000*. Cambridge, MA/London, England: Harvard University Press.

Milleville, P. (2007). *Une agronomie à l'œuvre. Pratiques paysannes dans les campagnes du Sud*. Paris: Arguments -QUAE.

Moriceau, J. M. (1999). *La terre et les paysans aux XVII et XVIIIè siècles, Guide d'histoire agraire,* . Rennes: Bibliothèque d'Histoire Rurale 3, Association d'Histoire des Sociétés Rurales.

Morlon P. (coordinateur) (1992). *Comprendre l'agriculture paysanne dans les Andes Centrales, Pérou – Bolivie*. Paris: INRA éditions.

Olivier de Sardan, J. P. (1994). De l'amalgame entre analyse-système, recherche participative et recherche-action, et de quelques problèmes autour de chacun de ces termes. *Recherches-système en agriculture et développement rural, Symposium international*, Montpellier, France, nov. 1994, CIRAD 1996, pp. 129–140.

Olivier de Sardan, J. P. (1995). *Anthropologie et développement. Essai en socio-anthropologie du changement social*. Paris: APAD-Karthala.

ORSTOM. (1985). *A travers champs, agronomes et géographes*. Paris: Editions de l'ORSTOM, coll. Colloque et séminaires.

Ostrom, E. (1990). *Governing the commons. The evolution of institutions for collective action*. Cambridge: Cambridge University Press.

Osty, P. L. (1978). L'exploitation agricole vue comme un système. Diffusion de l'innovation et contribution au développement". *Bull. tech. Inf. Min. Agr., 326*, 43–49.

Paul, J. L., et al. (1994). Quel système de référence pour la prise en compte de la rationalité de l'agriculteur: du système de production agricole au système d'activité. *Les Cahiers de la Recherche-développement, 39*, 7–19.

Peet, R., & Watts, M. (Eds.). (2004). *Liberation ecologies. Environment, development, social movements* (2nd ed.). London/New York: Routledge.

Pélissier, P. (1966). *Les Paysans du Sénégal. Les civilisations agraires du Cayor à la Casamance*. Saint-Yrieix (Haute-Vienne): Imprimerie Fabrègue.

Pélissier, P. (1979). Le paysan et le technicien. Quelques aspects d'un difficile face-à-face. In *Maîtrise de l'espace agraire et développement en Afrique tropicale* (pp. 1–8). Actes du Colloque de Ouagadougou (4–8 décembre 1978). Paris: ORSTOM.

Pélissier, P., & Raison, J. P. (2001). *"Préface", préface de l'ouvrage de Hubert Cochet: Crises et révolutions agricoles au Burundi* (pp. 7–16). Paris: Karthala/INA P-G.

Pépin-Lehalleur, M., & Sautter, G. (1988, déc.). Mante (Tamaulipas, Mexique): un système agraire régional ?, *Les Cahiers de la Recherche-Développement, 20*, 17–28, CIRAD.

Perrot, C., & Landais, E. (1993). Exploitations agricoles: pourquoi poursuivre la recherche sur les méthodes typologiques. *Les cahiers de la Recherche Développement, 33*, 13–23.

Pillot, D. (1987). *Recherche Développement et Farming System Research, Concèpts, approches et méthodes* (40 p.). Réseau Recherche-Développement, Ministère de la Coopération, Paris.

Pillot, D. (1992, juin). Je sais avec qui je suis en désaccord, mais je cherche toujours avec qui je suis en accord, réflexion sur la diversité des approches systémiques du milieu rural. *Colloquio mesoamericano Sistema de producción y desarollo agrícola*. Mexico.

Raison, J. P. (1984). *Les Hautes Terres de Madagascar*. Paris: ORSTOM/Karthala.

Rastoin, J. L. (1996, décembre). Dynamique du système alimentaire français. *Agroalimentaria, 3*.

Rastoin, J. L. (2008). Les multinationales dans le système alimentaire. *Projet, 307*, 61–69.

Reboul, C. (1977). Déterminants sociaux de la fertilité des sols. Fertilité agronomique et fertilité économique". *Actes de la Recherche en Sciences Sociales, 17–18*, 88–112.

Reboul, C. (1989). *Monsieur le capital et Madame la terre, fertilité agronomique et fertilité économique*. Paris: EDI/INRA.

Renard, J. (2002). *Les mutations des campagnes, paysage et structures agraires dans le monde*. Paris: A. Colin.

Richards, P. (1985). *Indigenous Agricultural Revolution. Ecology and food production in West Africa*. London: Hutchinson.

Risler, E. (1897). *Géologie Agricole*. Nancy: Imprimerie Berger-Levrault et Cie.

Rodgers, E. M. (1962). *Diffusion of innovations*. New York: Free Press.

Ruf, F. (1995). *Booms et crises du cacao, les vertiges de l'or brun*. Paris: Ministère de la coopération – CIRAD-SAR – Karthala.

Ruthenberg, H. (1980). *Farming systems in the tropics* (3rd ed., 366 p). Oxford: Oxford Science Publications, Clarendon Press.

Sacklokham, S., & Dufumier, M. (2006). Land tenure policy, deforestation and agricultural development in Lao PDR: the case of the Vientian plain. *Moussons, 9–10*, 189–207, *Marseille*.

Sautter, G. (1985). Paysagismes. In C. Blanc-Pamard & A. Lericollais (Eds.), *A travers champs agronomes et géographes, Dynamique des systèmes agraires* (pp. 289–297). ORSTOM Editions, coll. Colloques et Séminaires.

Sautter, G. (1993). *Parcours d'un géographe. Des paysages aux ethnies, de la brousse à la ville, de l'Afrique au Monde*. Paris: Editions Arguments (2 tomes).

Sautter, G., & Pélissier, P. (1964) Pour un atlas des terroirs africains: structure-type d'une étude de terroir. *L'Homme, IV*(1), 56–72, Paris.

Sauvy, A. (1958). *De Malthus à Mao Tsé-Toung, Le problème de la population dans le monde*. Paris: Editions Denoël.

Scott, J. C. (1990). *Domination and the arts of resistance: hidden transcripts*. New Haven: Yale University Press.

Scott, J. C., & Bhatt, N. (Eds.). (2001). *Agrarian studies: Synthetic work at the cutting edge* (p. 320). New Haven/London: Yale University Press.

Sébillotte, M. (1974). Agronomie et agriculture. Essai d'analyse des tâches de l'agronome. *Cahiers de l'ORSTOM, Série Biologie, 3*(1), 3–25.

Sébillotte, M. (1976). *Jachère, système de culture, système de production*. Paris: Institut National Agronomique Paris Grignon.

Sébillotte, M. (1989). Fertilité et systèmes de production, Essai de problématique générale". In M. Sébillotte (Ed.), *Fertilité et systèmes de production* (pp. 13–57). Paris: INRA.

Sébillotte, M. (1996). *Recherches-système en agriculture et développement rural*, actes du Symposium International (Montpellier, nov. 1994).

Seignobos, C. (1984). *Les instruments aratoires en Afrique tropicale, la fonction et le signe, Cahier ORSTOM série sciences humaines, XX*(3–4).

Seignobos, C. (2000). *Outils aratoires en Afrique, Innovations, normes et traces*. Paris: Karthala/IRD.

Shanin, T. (Ed.). (1970). *Peasants and peasant societies*. London: Penguin.

Sigaut, F. (1975). *L'agriculture et le feu, Rôle et place du feu dans les techniques de préparation du champ de l'ancienne agriculture européenne*. Paris: Mouton & Co.

Sigaut, F. (1976a). Une discipline scientifique à développer: La technologie de l'Agriculture (1). *Cah. Ing. Agronomes, 307*, 16–21.

Sigaut, F. (1976b). Une discipline scientifique à développer: La technologie de l'Agriculture (2). *Cah. Ing. Agronomes, 309*, 27–30.

Sigaut, F. (1981). Pourquoi les géographes s'intéressent-ils à peu près à tout sauf aux techniques ? *L'Espace Géographique, 4*, 291–293.

Sigaut, F. (2004, juillet-août-septembre). L'agriculture comparée réinventée. *Ingénieur de la vie, 468*, 17–18.

Sumberg J., & Thompson, J. (2010). Contested agronomy: Pegs, plots & politics in Sub-Saharan Africa, IDS, May 2010 (projet).

Tiffen, M., Mortimore, M., & Gichuki, F. (1994). *More people, less erosion, environmental recovery in Kenya* (p. 312). Nairobi: ACTS Press (in association with the Overseas Development Institute, London, UK).

Trochet, J. R. (1993). *Aux origines de la France rurale. Outils, pays et paysages*. Paris: Mémoires et Documents de Géographie, CNRS Editions.

Verdeaux, F. (1997). Quand la campagne était une "forêt vierge... L'invention de la ruralité en Côte d'Ivoire - 1911-199.... In J. M. Gastellu & J. Y. Marchal (Eds.), *La ruralité dans les pays du Sud à la fin du XXè siècle* (pp. 79–97). Paris: Editions de l'ORSTOM.

Viallon, J. B. (1985). Sources et méthodes utilisées. In D. Aubert et al. (Eds.), *Systèmes de production et transformations de l'agriculture 1. Essai de bilan des travaux du département d'économie et sociologie rurale*. INRA.

Vissac, B. (1979). *Eléments pour une problématique de recherche sur les systèmes agraires et le développement*. Toulouse: INRA-SAD.

Young, A. (1792). *Travels during the years 1787, 1788, & 1789: undertaken more particularly with a view of ascertaining the cultivation, wealth, resources, and national prosperity of the Kingdom of France*. London: Bury St. Edmund's.

MIX
Papier aus verantwortungsvollen Quellen
Paper from responsible sources
FSC® C105338

If you have any concerns about our products,
you can contact us on
ProductSafety@springernature.com

In case Publisher is established outside the EU,
the EU authorized representative is:
Springer Nature Customer Service Center GmbH
Europaplatz 3, 69115 Heidelberg, Germany

Printed by Libri Plureos GmbH
in Hamburg, Germany